Energy Efficiency and Conservation in the Developing World

The World Bank's Role

A World Bank Policy Paper

The World Bank
Washington, D.C.

1818 H Street, N.W.
Washington, D.C. 20433 U.S.A.

Manufactured in the United States of America
First printing January 1993

Library of Congress Cataloging-in-Publication Data

Energy efficiency and conservation in the developing world : the World Bank's role.
p. cm. — (A World Bank policy paper, ISSN 1014-8124)
Includes bibliographical references.
ISBN 0-8213-2317-2
1. Energy consumption—Developing countries. 2. Energy conservation—Developing countries. 3. Energy policy—Developing countries. I. International Bank for Reconstruction and Development. II. Series.
HD9502.D442E522 1993
333.79'09172'4—dc20 92-46611
CIP

Cover design by Walton Rosenquist and Beni Chibber-Rao

ISBN 0-8213-2317-2
ISSN 1014-8124

Contents

Figures

Acknowledgments

The principal authors of this paper are Robert J. Saunders and Sunita Gandhi. Anthony Churchill contributed to the development of the policy and analytical framework and to the paper's overall structure. Peter Cordukes, John Besant-Jones, Rangaswami Vedavalli, and Dennis Anderson contributed background material and critical review. The paper also benefited enormously from extensive consultations with Bank operations and policy staff and with others outside the Bank who are too numerous to list.

Glossary

BOO/BOOT schemes

Build-Own-Operate (BOO) and Build-Own-Operate-Transfer (BOOT) schemes are methods by which private sector participation in the power sector is encouraged. Under these approaches, a project company under private ownership, or a joint venture with a minority public participation, is set up to plan, finance under limited recourse, design, construct, and operate power generation facilities. In a BOOT arrangement, ownership of the facility is ultimately transferred to another entity after a specified period of operation.

Country commitment

Commitment must be judged on a country-by-country basis within the framework of a country-assistance strategy around the themes of significant progress toward needed reform and no more "business as usual."

Demand-side management

Identifying and implementing initiatives that improve the use of energy-supply capacity by altering the characteristics of the demand for energy. DSM involves a mix of pricing, other load management, and conservation strategies designed to increase the incentives for a more efficient use of energy

Energy-efficiency improvements	Any measure that results in the delivery of any energy service with a reduction in energy consumption. Thus, carrier substitution or fuel-switching measures that lead to reductions in energy demand also become examples of energy-efficiency improvements.
Energy end use	Energy applications such as motive power, lighting, process heat, water heating, refrigeration, air cooling, cooking, and so on.
Integrated energy strategy	An interrelated set of measures that points the energy sector toward the most efficient, equitable, and environmentally-benign resource use. The strategy requires decisions on both the energy supply and demand side about sector structure, institutions, ownership, financing, fuel availability (coal, oil, gas), technology availability (import restrictions), structures of end-use markets, pricing policy, standards, service levels, and so on.
Integrated energy resource planning (IERP)	Primarily a U.S.-type planning process whereby utilities (and in the U.S., their regulatory commissions) evaluate available demand- and supply side-options to provide energy services (including purchased power) and determine an optimal energy service strategy, given economic and environmental factors. The essential concept of IERP is the equal treatment, or integration, of energy-based and conservation-based energy services. Planners attempt to rank by cost all the different energy supply and end-use technologies, processes, and programs that might be used to provide energy services and implement them beginning with the lowest-cost opportunities.
Load factor	A key measurement that compares a utility's average kilowatt-hour load to its peak, or maximum hour's usage, in a given year. A high load factor means greater plant utilization, since a company must build capacity to meet its peak demand, not its average demand.

Load management	Any effort to control loads by economic incentives, direct interventions, or new technology. Shifting load from peaks to valleys, or simply shaving the peak, defers capacity additions and transfers load from high cost, inefficient peaking generation to more economically efficient base-load units.
Marginal cost	The increase in the total costs of an enterprise caused by increasing its output by one extra unit. Marginal cost pricing is the setting of the price of an item equal to the cost of producing one extra unit of the item. Marginal cost represents the opportunity cost, or the total sacrifice to society, for producing an item. Long-run marginal cost is the cost of meeting an increase in consumption, sustained indefinitely into the future, when needed capacity adjustments are possible. In the long run, an increase in demand will result in a corresponding increase in the operating costs as well as in the capacity costs.
Peak pricing	The setting of higher prices than average when supplying services during a period of peak demand. Enough electricity capacity must be installed to satisfy demand at peak times, because, in general, electricity cannot be stored. At off-peak times the cost of electricity is lower at the margin than at the peak, at which less-efficient power stations have to be switched in to meet the demand.
Regulation	The supervision and control of the economic activities of private and arms-length public enterprises by government in the interest of economic efficiency, fairness, health, and safety. Regulation may be imposed simply by enacting laws and leaving their supervision to the normal processes of the law, by setting up special regulatory agencies, or by encouraging self-regulation by recognizing, and in some cases delegating powers to, voluntary bodies.

Executive Summary

There is a congruence of several forces in the developing world that makes very timely the formulation of a strategy to address energy efficiency and conservation issues better.[1] Many countries are becoming receptive to reforming the way energy is produced and consumed as they experience (a) rapidly growing demand for energy; (b) major constraints on available energy financing; (c) increased pressures to sustain the environment; (d) poor energy sector performance and unsatisfied customers; and (e) a reappraisal of the roles of government and of the public and private sectors in development. These factors are forcing developing countries to address long-neglected issues of energy wastage in production and end use. The World Bank now has an enhanced opportunity to further assist developing countries to improve their energy efficiency and conservation policies and practices.

Comparing the experience of developing and developed country performance in energy efficiency has highlighted four critical factors that directly correlate with differences in the efficiency of energy production and end use. These factors relate to:

- differences in energy pricing policies,
- mechanisms for controlling or regulating energy supply enterprises,
- the extent to which energy-using industries are protected from competition, and
- other legal, institutional, and information barriers to the efficient functioning of markets.

In formulating integrated energy strategies, governments must review both supply- and demand-side options and set priorities that address major impediments to improving the efficiency with which energy is produced and consumed. In many developing countries the first step in the formation of an integrated energy strategy will be to target demand-side issues such as energy pricing, promote competitive markets in which energy consumers must operate, and take selected information-related interventions to help markets work more efficiently. On the supply side, the highest priorities will be to make supply-side institutions responsive through institutional and regulatory reform and to make investments in efficiency-enhancing activities, such as plant rehabilitations and reducing transmission and distribution losses. Other priority areas, such as facilitating the ease of technology transfer and increasing the focus on energy efficiency in transport, involve initiatives on both the supply and the demand sides. In general, an increased focus is needed on providing energy services, including improved efficiency, as well as increasing supplies.

After reviewing World Bank and country experience, carrying out a comprehensive survey of the literature on energy efficiency and conservation issues, and undertaking extensive consultations with outside developing and developed-country officials, academics, researchers, practitioners, and nongovernmental organizations (NGOs), it is concluded that continuing efforts to address countrywide policy and institutional issues are the most important means by which to achieve the largest energy efficiency gains in developing countries. As a result, the main elements of existing Bank policy for achieving energy efficiency in the developing world will remain in force while the Bank fully exploits every additional opportunity to improve energy efficiency.

The Bank will continue its efforts toward increased lending for components designed to improve energy efficiency and promote economically-justified fuel switching. In addition, to take advantage of the increased receptivity of many developing countries to efficiency issues, the Bank will sharpen its focus on energy efficiency by undertaking the following four point program:

Point One. To gain greater country commitment, the Bank will better integrate energy efficiency issues into its country policy dialogue so that they can be addressed at an earlier stage.

In the Banks general country policy dialogue with developing countries, greater emphasis will be given to energy pricing and to fundamental institutional and structural factors that affect supply- and demand-side energy efficiency. The Bank will assist countries in putting viable integrated energy strategies in place. The energy sector is a candidate for greater attention because of its size, its strategic role in the growth process, and its major environmental impacts.

Point Two. The Bank will be more selective in lending to energy-supply enterprises.

Governments should clearly demonstrate that they are putting in place structural incentives that will lead to more efficient energy supply and consumption. The Bank will not continue to finance energy supply projects where poorly performing public energy enterprises and their governments are unwilling to carry out fundamental structural reforms that could significantly improve the ways they do business.

Point Three. Approaches for addressing demand-side management (DSM) and end-use energy intermediation issues will be identified, supported, and given high-level in-country visibility.

The Bank will increase its efforts to improve intermediation in the energy and industry information markets in developing countries in order to reduce the relatively high information, management, technology, and financing transactions costs. There is a role for both the public and private sectors. As the gap between the cost of energy and the price at which it is sold is reduced or eliminated, market intermediaries will increasingly be able to earn a profit through arbitration of information, technology, financing, and management assistance. The Bank will play a role by identifying, supporting, and financing both public and private sector institutional homes and initiatives that can serve intermediation functions and pursue DSM objectives.

Point Four. The Bank will give greater attention to the transfer of more energy-efficient and pollution-reducing technologies in its sector and project work.

For all sectors, including basic materials processing industries, the Bank will actively monitor, review, and disseminate the experience of new efficiency-enhancing supply-side and end-use products, technologies, and processes, and pollution-abating technologies as they are developed and reach the marketplace; help finance their application; and encourage the reduction of barriers to their adoption. Staff working in all sectors will explicitly review technology choice options during project appraisals and in sector work.

The importance of such an intensified effort to promote energy efficiency was underscored at the United Nations Conference on Environmental Development (UNCED) in June 1992 by the commitment of the international community, at the highest levels, to energy efficiency and other cost-effective greenhouse gas control measures.

1

Introduction and Overview

The demand for energy is growing rapidly in developing countries. But energy is produced and consumed with varying low levels of efficiency. In the early 1980s the Bank began to promote new kinds of projects with physical components and policy initiatives that specifically targeted energy efficiency and conservation and economically justified fuel switching. Other donors, and some countries, have had similar programs, and Bank cooperation with them has been extensive.

Given these efforts, the efficiency with which energy is produced and consumed in developing countries would be expected to have improved greatly. In a few countries—Korea, Malaysia, Chile—that has been the case. In many other developing countries, however, success has been limited; energy supply and end-use efficiencies are still only two-thirds to one-half of what would be considered *best practice* in the developed world. Studies indicate that at current relatively low energy prices and with the present state of technology, a saving of 20 to 25 percent of energy consumed could be achieved economically in many developing countries with existing capital stock. If investments were made in new, more energy-efficient capital equipment, a saving on the order of 30 to 60 percent would be possible.

In light of these possibilities, it is reasonable to ask why energy efficiency in the developing world remains so low. Has the World Bank done enough of the right things to encourage the better use of energy resources? How can the Bank's programs be strengthened?

After reviewing Bank and country experience, carrying out a comprehensive survey of the literature on energy efficiency and conservation issues (see bibliography), and undertaking extensive consultations with outside developing- and developed-country officials, academics, researchers, practitioners, and NGOs, this paper concludes that continuing efforts to address countrywide policy and institutional issues are the most important means by which to achieve the largest energy efficiency gains in developing countries. As a result, the main elements of existing Bank policy for achieving energy efficiency in the developing world will remain in force while every additional opportunity to improve energy efficiency is fully exploited. These existing elements are discussed as follows.

Energy prices

Removing subsidies for energy production and use is the first priority. According to *World Development Report 1992*, subsidies now amount to more than $150 billion[2] per year in developing countries—for electricity consumption alone the subsidy is on the order of $100 billion per year. Such subsidies waste capital and energy resources on a very large scale.

Subsidizing the price of electricity is both economically and environmentally inefficient. Low prices give rise to excessive demands and, by undermining the revenue base, reduce the ability of utilities to provide and maintain supplies. Developing countries use about 20 percent more electricity than they would if consumers paid the true marginal cost of supply. Underpricing electricity also discourages investment in new, cleaner technologies and more energy efficient processes.

In recent years a number of organizations have proposed that, in investment planning, the least-cost concept should be extended to include demand-side planning. The Bank has had a longstanding goal of bringing about *least-cost demand-side efficiency*, principally through price and institutional reforms and policies that create competition in end-use markets. When energy is highly subsidized, it is difficult for many demand-side initiatives that might reduce energy use to compete. The purpose of good pricing policies is to convey to consumers the real costs of energy, including the real costs of pollution, so that cost-effective consumption decisions are made from a social perspective. From both demand-side and supply-side perspectives, good pricing policies are fundamental prerequisites for achieving energy efficiency.

Supply-side Institutional Reform

Inappropriate investment decisions and government intervention in the day-to-day operations of energy-supply enterprises result in large economic costs and financial losses. Increased emphasis on supply-side restructuring will, therefore, aim at making energy-supply enterprises more autonomous and decisionmaking more transparent. The improved managerial efficiency that would arise from institutional reform

Box 1. The Meaning of Energy Efficiency

Energy efficiency is not a well-defined concept. Traditionally it has had two aspects: managerial, or "X"-efficiency, and price efficiency. More recent literature suggests that "X"-inefficiencies exist not only in production but also on the energy consumption side. One frequently-used empirical measure of energy efficiency is energy use per unit of GDP (see Box 5). However, in a development context, it is the economic rate of return to capital that is relevant, not energy use per unit of output of GDP. If lower energy consumption per unit of GDP were achieved by reducing the rate of return to investment for the reasons discussed in this paper—that is, increasing subsidies and raising technical/managerial/"X"-inefficiencies—this would not be desirable. Pursuing increased energy efficiency for its own sake could be costly; it is only worth pursuing up to the point where the change in the economic rate of return becomes zero or negative (or alternatively, where the economic rate of return is greater than or equal to the cost of capital).

and greater private sector participation is also dependent on achieving good financial rates of return that would leave the enterprise financially able to undertake required maintenance and reduce physical losses. The longstanding requirement that Bank-financed projects and programs be part of least-cost investment plans to provide energy services is fully consistent with the goals of producing and using energy efficiently. The least-cost requirement should include full compliance with environmental policies.

Competitive Markets

The protection of domestic industry from competition has been a significant factor in the relatively high energy consumption per unit of output observed in many developing countries. The World Bank will continue to pursue vigorously policies that will bring about competitive end-use markets in developing countries.

There is also scope for two further Bank initiatives to improve energy efficiency in developing countries through:

Market intermediation

An increased focus on improved intermediation in the energy and industry information markets in developing countries is needed to reduce the relatively high information, management, technology, and financing transaction costs. Even in OECD countries, where energy prices come closer to reflecting costs and where most end-use markets are competitive, it has been demonstrated that information programs help narrow the gap between the the technical potential for energy efficiency and current efficiency levels. These programs provide producers and consumers with examples and with the demand-side technical, economic, and financial information they need to make decisions on energy production and consumption. Intermediation functions will be identified and supported in all developing countries.

Technology transfer

Many developing countries have a disproportionately high level of old, inefficient, and polluting capital stock in all sectors. There is great scope for introducing through Bank operations new efficiency-enhancing and pollution-abating supply-side and end-use products, technologies, and processes.

This paper provides a background on the evolution of developed versus developing-country energy efficiency and summarizes World Bank experience (chapter 2). It outlines the growing energy demand in developing countries and the accompanying financial and environmental problems that bring energy-efficiency issues to the forefront (chapter 3). The paper then sets out the potential for improved energy efficiency

Box 2. Power Loss Reduction

Unaccounted energy losses were reduced significantly by improving power sector management practices under the Afghanistan First Power Project. During the period 1974-79, the power utility, DABM, was able to increase its sales by about 62 percent with only a 19 percent increase in generation. DABM also improved its financial position significantly with only a 20 percent increase in tariffs over the five-year period. To reduce energy losses, the government and DABM initiated vigorous measures that included intensifying inspection of customers' premises; disconnecting illegal and nonpaying customers; discharging or relocating staff to disrupt vested interests; increasing and more tightly controlling meter reading and testing; implementing faster billing; and prosecuting consumers found guilty of stealing power.

Source: Gandhi *et al* 1993.

and the main reasons that efficiency has been so poor (chapters 3 and 4). Finally, it summarizes the priority actions for achieving energy efficiency in developing countries and sets out a four-point program for the World Bank (chapters 5, 6, 7).

This paper does not specifically address issues related to the efficiency with which traditional fuels—straw, dung, wood, and charcoal—are gathered or produced and consumed. While many of the issues in the traditional energy sector are the same as those for modern fuels—including better management of supplies, interfuel substitution, energy demand management, and pricing—the solutions are different. The pricing issues in this paper focus on how to include the value of externalities in the market price rather than on simply whether to eliminate government subsidies. Improving the efficiency of traditional supplies extends beyond the commercial energy sector to a number of sectors, including agriculture, forestry, and traditional fuels. The efficiency and conservation issues for traditional fuels will be examined in a subsequent paper on rural energy.

2

Evolution of Energy Efficiency

By the early 1970s, most developed countries had experienced decades of low energy prices and plentiful fuel supplies, with a consequent high and growing per capita use of energy. This high usage was of little concern to most governments until the first oil shock, when rapidly rising energy prices and interruptions in supplies forced reexamination of existing policies. In most developed countries, conservation and end-use efficiency improvements became an important component of energy policy. Fortunately, there was ample scope for improvements. Insulation and other measures were applied to reduce space heating and air conditioning requirements. The wide use of household appliances meant a large market for more efficient electric motors, compressors, and other equipment. In transportation, existing technologies were applied to the production of new vehicles, with consequent improvements in fuel use.

Also in the developed countries, the growing demand for many energy-using consumer and producer goods, as record numbers of young people entered the age of household formation, was met in part through consumer and producer goods that incorporated newer, more energy-efficient technologies. By the mid-1980s, a large part of the capital stock of these goods consisted of household appliances and industrial motors that reflected these newer technologies. The trend toward efficient technologies is shown clearly in energy consumption statistics; as the level of services rose, overall per capita energy demands decreased. Although in recent years the rate of improvement has slowed as energy prices have dropped back to historically lower levels, the experience of the 1970s and early 1980s was dramatic proof of the potential for improving energy efficiency.

A similar improvement in energy efficiency has not occurred in most of the developing world. On the contrary, in many instances energy intensity (energy consumed per unit of output) has continued to increase. Four main factors account for this disappointing performance. First, many governments failed to pass on all of the increase in international energy prices to domestic producers and consumers. Second, in many countries a large part of total consumption was and is in the nonhousehold sectors, which are dominated by inefficient state enterprises and protected industries. Third, the subsidized publicly-owned monopoly enterprises supplying energy in cost-plus conditions did not provide a conducive environment for effective built-in incentives for high levels of efficiency. Fourth, because of noncompetitive market structures and subsidized energy prices, it has not been profitable for market intermediaries to develop to arbitrate information on energy efficiency or on financial and technology options.

In the developing world the continued rapid growth in energy demand and its financial and environmental consequences have stimulated renewed interest in exploring the potential to improve efficiency in both supply and end use. A number of technical studies have shown that if developing countries were to use the best practices and technology now available, dramatic declines in new energy requirements would theoretically be possible. This technical potential for greatly improving energy efficiency has attracted much attention from both environmentalists and energy planners. Financial development agencies such as the World Bank have also been looking for ways to give greater weight to

Box 3. Improving Power Sector Efficiency With Advanced Lighting Units

The Bank's 1990 Electricity Transmission and Conservation Project in Brazil aimed to increase public lighting efficiency by replacing 175-watt incandescent bulbs with 80-watt fluorescent lamps and 50-watt high-pressure sodium fluorescent lamps. Brazil has 770,000 public lighting points that operate with incandescent bulbs. The program was expected to save about 103 GWh/year, enabling the government to defer installation of 27 MW of additional capacity, with potential capital cost savings of $40 million. Unfortunately, the project was recently canceled because of Brazil's failure to meet the conditionality of the electricity tariff agreement.

Source: Gandhi *et al* 1993.

Box 4. Energy Efficiency Projects

Since 1979 the Bank has funded more than forty projects entirely devoted to energy efficiency. Many of these projects have directly instituted technical changes and improvements, rehabilitated power and industrial facilities, promoted conservation and household measures in all sectors, and promoted demand and load management in the power sector. At least eleven projects have targeted energy efficiency improvements in petroleum refineries, which are the most energy-intensive operations in many countries. Projects have also included efficiency improvements of manufacturing plants in the fertilizer (Yugoslavia, Turkey, Indonesia, and China), pulp and paper (Turkey), textiles (Philippines, Turkey), cement (India), and metals (Egypt and Guyana) industries. The majority of these projects also helped establish or strengthen energy-efficiency institutions and energy-related physical, economic, and financial infrastructure. Institutional capability played a significant role in the effectiveness of project implementation in many cases. Lack of institutional consistency and follow-through contributed to project failure in several cases (Guinea, Jamaica, Liberia, Sudan, and Uruguay).

Source: Gandhi *et al* 1993.

energy conservation and efficiency in their policy dialogue and investment decisions.

In fact, this potential was one of the driving forces behind the Bank's establishment of the Energy Sector Management Assistance Program (ESMAP) in 1983 in cooperation with UNDP and the donor countries. It was also behind Bank efforts in the energy sector increasingly to complement new energy supplies with programs or investments specifically designed to increase the efficiency with which existing capital stock and new plants produce and consume energy. The Bank's 1983 Electric Power Sector Support Strategy Paper explicitly recommended that "tariff analysis should emphasize economic efficiency as well as financial viability" and that "project identification and preparation work should pursue opportunities for new-style projects including maintenance and rehabilitation, conservation, and efficiency."

World Bank Experience with Energy Efficiency

A preliminary review of some 1,500 World Bank Staff Appraisal Reports (SARs) was undertaken as background for this paper to identify projects with explicit energy efficiency components. From that number, 233 projects were selected for further analysis using the narrow guidelines summarized in the appendix. For these, there were sixty-one Project Completion Reports (PCRs) available. Once the projects were selected, they were organized and analyzed in two ways. First, projects were reviewed to determine the extent to which they targeted energy efficiency in specific areas:

- technological reform,
- interfuel substitution,
- institutional development,
- demand-side management, and
- energy policy and regulatory reform issues.

Projects with energy efficiency components were also reviewed by sector (industry, energy, power, transport, and fossil fuels) and according to two other categories (structural adjustment loans and projects that were completely dedicated to energy efficiency). Second, site visits were made to ten countries to examine specific issues relating to the structure and mandate of energy efficiency institutions.

The energy-efficiency impacts of Bank-induced policy or institutional reform can only be assessed in broad and general terms. It is possible, however, to assemble information on the level of Bank activity in putting together specific projects or components of projects that directly address energy-efficiency issues. In doing this, it was found that Bank lending for components designed to increase energy efficiency and promote economically justified fuel switching has increased significantly in recent years (figure 1). The power sector has led the way, with efficiency components comprising slightly more than 50 percent of the total components identified. Loans for the energy, industry, and transport sectors each had between 8 and 14 percent of all components identified. Of the areas reviewed, structural adjustment lending had the fewest identifiable energy-efficiency components, with only 3 percent (figure 2).

In the power sector, in addition to promoting policy, pricing, and institutional reform, the Bank has increasingly pushed programs to rehabilitate old plants to extend their lives and increase productive efficiency; strengthen and upgrade transmission and distribution systems to reduce technical losses; improve commercial practices to reduce nontechnical losses; promote least-cost system planning and operations to supply power from the most cost-efficient plants; establish energy audits to reduce wastage; promote technology upgrading; and set up dedicated energy-efficiency institutions.

Likewise, in the oil and gas sectors, most recent projects have contained combinations of energy-efficiency components similar to those in the power projects, covering loss reduction; load management; plant operational improvements; better operating procedures; process modi-

Figure 1. Financing for Energy Efficiency Components in Projects Partly Financed by the World Bank

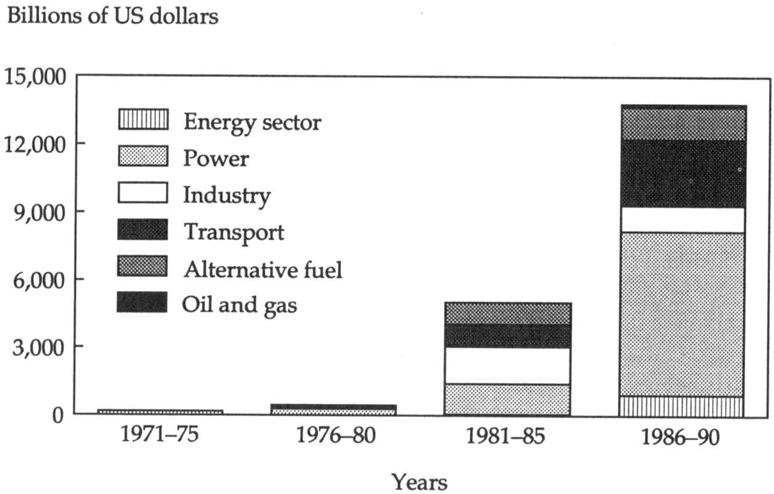

Source: Gandhi *et al* 1993.

Figure 2. Percentage of Projects with Energy Efficiency Components (1968–91)

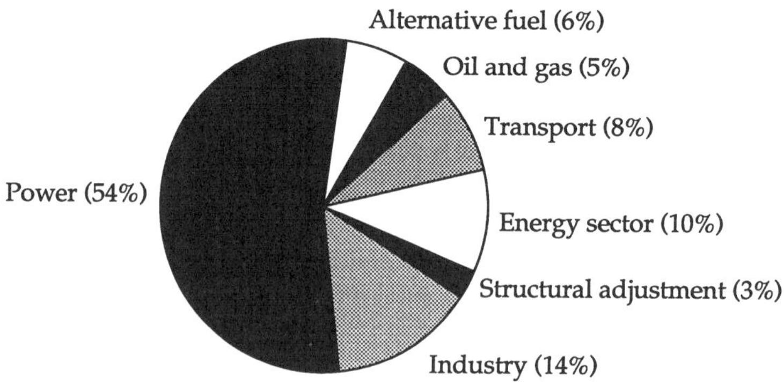

Source: Gandhi *et al* 1993.

fication; energy-enterprise management; and institutional strengthening. Bank refinery projects have been almost solely focused on rehabilitation to achieve both operational and energy efficiency rather than on financing expansion or new facilities.

The Bank also has promoted natural gas as an energy-efficient fuel. Methods for increasing management accountability in the oil and gas sectors, and for encouraging private sector participation, have been a focus of the Bank's oil and gas program, just as they have been in the power sector. Similar initiatives were also carried out in the industry sector, such as loans for plant rehabilitations, technology retrofitting, energy audits, and the reduction of market barriers to competition.

The lessons learned from the review of Bank experience are many and varied. Briefly, however, it is clear that while many individual efficiency components have been successful, at least in the short run, many other economically justified efficiency-related initiatives on both the energy supply and demand sides were not undertaken—boiler tuning, better plant housekeeping, and more efficient motors, light bulbs, and appliances—or did not realize their full potential. The reasons for this failure, in many instances, have been found to relate directly to one or more of the following factors:

- lack of government follow-through on agreed commitments to efficiency-related policy and institutional change,
- weak or absent competitive market forces in the commercial and industrial sectors,
- subsidized fuel and electricity prices,
- administered trade barriers and discriminatory taxes and subsidies,
- lack of local financing at competitive rates,
- lack of in-country technical knowledge of energy-efficiency options,
- lack of indigenous industries for supply of energy-efficient products and efficiency services,
- the relatively high weight given to first-cost considerations when making equipment purchases, and
- absence of specific incentives for increasing efficiency, such as mandated energy performance codes and standards for industry, transport, and buildings.

3

Forces Driving Increased Energy Efficiency

Growing Energy Demand in Developing Countries

It is unavoidable that developing countries will have to increase the amount of commercial energy they consume in order to improve the welfare of their populations. The electric power consumption of the whole of Sub-Saharan Africa, for example, is less than that of New York City. Even Brazil consumes only about one-tenth as much energy per capita as the United States or Western Europe. On an overall per capita basis, developing countries annually consume about 0.4 toe of commercial fuels, or 0.6 toe including biofuels, compared with more than 3.2 toe in Western Europe and 7.4 toe in the United States. The difference in electricity use is particularly striking. On average, developing countries use only 500 kWh of electricity per capita per year, compared with more than 5000 kWh in Europe and more than 10,000 kWh in the U.S.

Given the relatively low consumption base, the growth in demand for commercial energy in the developing countries is expected to account for almost all of the increase in future world energy consumption. This demand-growth is driven by several related factors, including the growth of populations and per capita incomes; the migration to urban areas, which in many instances leads to substitution of commercial energy for fuelwood; the increasing penetration of energy-intensive products and technologies (fertilizers, petrochemicals, cement, vehicles, appliances, motors); and the poor efficiency with which energy is produced and consumed in the developing world.

On a global scale the rate of growth in energy consumption in the developing countries over the past two decades has been more than seven times that of the OECD countries (5.3 percent per year compared with 0.7 percent per year). Growth in the electric power sector has been

Figure 3. Energy Consumption Projections

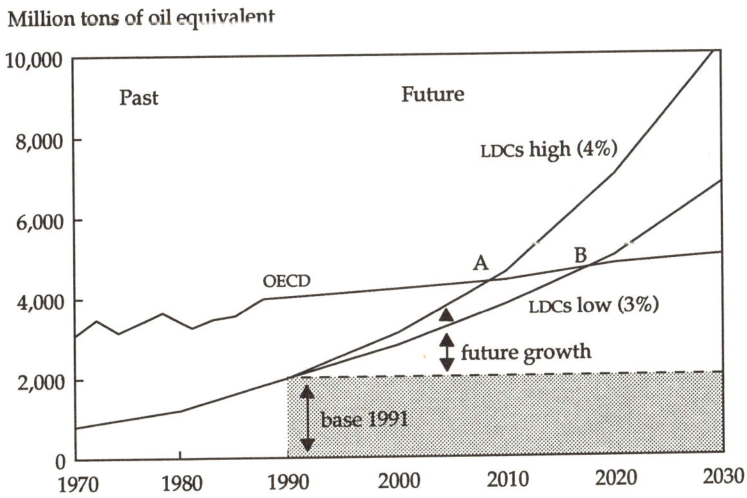

Source: Gandhi and Saunders 1993.

particularly dramatic. In more than 90 percent of a recent sample of fifty-one developing countries, installed capacity and generation per capita grew at more than twice the real growth rate of GDP. In 57 percent of those countries, installed capacity and generation grew at three times the real GDP growth rate. Power connections grew at 9 percent a year, or about two-and-a-half times the average population growth rate. Assuming 4 percent growth per year (two percentage points below a projected rate with full economic recovery in Latin America and Africa), developing country per capita energy consumption will still be less than one-quarter the level of OECD countries forty years from now. Nevertheless, given the high population and urban growth expected in developing countries, total commercial energy consumption in developing countries will likely be greater than in the OECD countries within fifteen to twenty years and more than four times that of Eastern Europe and the former USSR combined.

Financing Constraints

Given this rapidly growing demand, developing countries are beginning to experience increased pressures to produce and consume energy more efficiently, partly because of financing constraints on new energy supplies. Today in some developing countries, one-quarter to one-third of the public resources available for investment are going solely to electric power, and these investments are still inadequate. There are also major constraints on the availability of international concessionary and commercial bank finance, as well as on domestic public resources. This leaves only the domestic and international capital markets as potential sources of significant new investment capital. Domestic capital markets in many countries are weak, and it will take time to develop strong and viable markets. In the meantime, for electric power supply alone, developing countries want investments of $100 billion a year over the next decade. For Eastern European and Central Asian power utilities, it has been estimated that more than $70 billion a year between 1990 and 2000 will be required to bring them up to the technical standards that existed in 1990 in the former West Germany.

These magnitudes of funding have little chance of being mobilized. *World Development Report 1992* estimates that prices that are 50 percent of supply costs, combined with existing losses in capacity and energy from technical inefficiency, serve to double investment requirements in power supply in many countries. Removing such economic drains would alleviate some of the demand on available funding, but it is clear that more must be done. Producing and consuming energy more efficiently will have to become an increasingly important "source" of new energy supply.

Energy and the Environment

The increased focus on energy efficiency is also being driven by the fact that the environmental implications of the growth in energy demand in developing countries are very large worldwide. It is a fact that energy production and end use are major contributors to environmental degradation. The inefficient combustion of household fuels, for example, already shows up in serious respiratory and other health problems in low-income households, and poor practices in burning coal in power plants have already resulted in substantial deterioration in the quality of both air and water. Also, the poor management of urban congestion and automobile use is contributing to deteriorating air quality in many cities. On a global scale, the production and use of energy accounts for between 50 and 60 percent of the greenhouse trace gas emissions in the atmosphere. In general, energy efficiency is highly correlated with income levels. Poor households tend to be less energy efficient than higher-income households, and poor countries are less energy efficient than higher-income countries.

A number of studies of energy and the environment have concluded that environmental degradation can be reduced by switching to cleaner

Figure 4. Global Sectoral Contribution to Greenhouse Gases

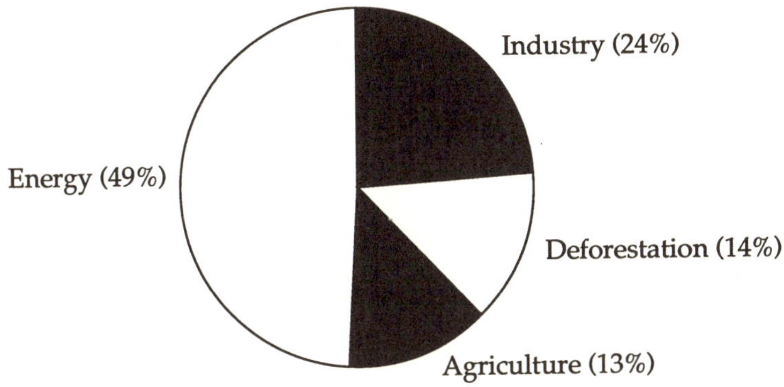

Source: World Bank data.

fuels and reducing the energy intensity of economic activity by using best practices and best technology in both energy production and consumption. This is not the complete picture, however. *World Development Report 1992* documents clearly that a twofold approach is required to reduce pollution from energy production and use. One part of the approach is to improve energy efficiency through cost-reflecting prices and policy and institutional reforms; this would also reduce costs and increase economic efficiency. The other is to use environmental taxes and regulations as an incentive for the energy industry and its consumers to adopt cleaner fuels (such as gas) and clean-fuel technologies (such as particulate emission controls and, where merited, scrubbers or advanced combustion technologies). Taken together, these two sets of policies are capable of reducing pollution dramatically while leading to cost savings in energy production and use.

Figure 5 illustrates this point for particulate emissions from power generation. Under a "business as usual" scenario, pollution would rise in line with demand growth. Raising prices to cost-reflecting levels, and instituting policy and institutional reforms, as discussed in this paper, would reduce pollution by slowing demand growth and thus reducing the amount of fuel needed. Furthermore, policies that encourage the control of particulate emissions directly can have dramatic effects on pollution abatement while adding relatively little to costs.

Similar conclusions apply to most other pollutants—e.g., to emissions from vehicles (figure 6). Controlling SO_2 and NO_x can be more expen-

Figure 5. Particulates Emissions from Electricity: Three Scenarios in Developing Countries

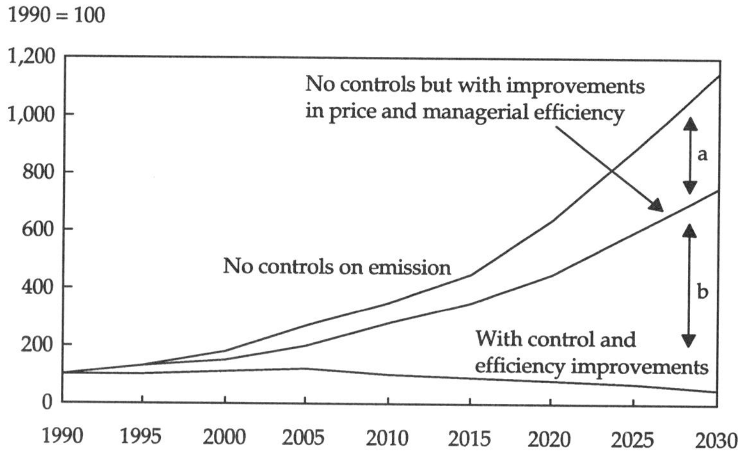

a) Contribution of energy efficiency
b) Contribution of low-polluting technologies

Source: *World Development Report 1992.*

sive, though the costs are not large relative to the gains from efficiency. A combination of economic efficiency and good pollution abatement policies is therefore very powerful and can improve the environment while raising incomes and social welfare. However, the developing world is only beginning to appreciate the seriousness of the pollution problems that result from poor management of the energy transformation process.

Potential for Greatly Improved Energy Efficiency in Developing Countries

Because of their current low efficiency base, developing countries have the potential to achieve significant energy efficiency gains. For example, in the petroleum sector, losses in refining (excluding the energy requirements of the refining process) are as high as 5 percent but could be as low as 0.5 percent in a well-run refinery—the difference represents mainly leakage and the flaring of refinery gases. Estimates of avoidable costs in the petroleum product supply chain in Sub-Saharan Africa—before the products even get to the consumer—amount to $1.3 billion per

Figure 6. Lead Emissions from Land Transport: Three Scenarios in Developing Countries

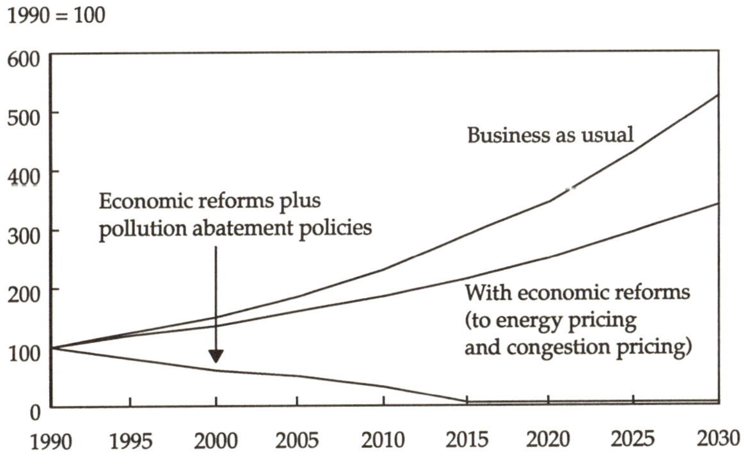

Source: World Development Report 1992.

year (about one-quarter of the entire Sub-Saharan petroleum import bill). Since petroleum product imports absorb one-third of export earnings in the region, this is a substantial sum exceeding total Bank adjustment lending to Africa for an average year.

In the power sector, which is always one of a developing country's largest commercial energy consumers, the problem is particularly acute. A recent review of the operational performance of the largest power utilities in fifty-one developing countries over a twenty-year period shows a general trend of declining technical efficiency. It has been estimated that older power plants in many developing countries consume from 18 to 44 percent more fuel per kilowatt hour of electricity produced than do plants in OECD countries, and they suffer transmission and distribution losses two to four times higher. In fact, technical and nontechnical transmission and distribution system losses in the delivery of electricity are commonly greater than 20 percent—occasionally approaching 40 percent.

Currently transmission and distribution losses represent about 31 percent of generation in Bangladesh, 28 percent in Pakistan, and 22 percent in Thailand and the Philippines. (In the United States only 8 percent of electricity is lost during transmission; in Japan, 7 percent.)

These losses, the equivalent of about 75,000 megawatts of capacity and 300 terawatt hours (300 billion kilowatt hours) a year, represent a loss to developing countries of approximately $30 billion a year through increased supply costs. Worse, by the end of the century, based on present trends, aggregate losses would double. It has been estimated that cutting transmission losses by only one-tenth in Asia would reduce the need for investment in generating capacity during the 1990s by about $8 billion—almost enough to pay for controls to reduce particulate emissions for every new power plant to be built in the entire developing world during the 1990s. While some losses represent theft and shortcomings in billing and collection, it is clear that technical losses in networks are very high.

The economic impacts of policies and investments that improve energy efficiency in developing countries can be substantial, first, because

Box 5. Energy Intensity

The ratio of energy input to GDP output is often used as a measure of the differences in energy efficiency among countries. However, it has been well-documented that this comparison can be a misleading indicator of overall energy efficiency performance. In making cross-country comparisons, there are problems with both the GDP numbers (exchange rates, omissions in income accounts) and with energy numbers (exclusion of biomass fuels). In fact, given the low prices for energy in many developing countries, in some cases it may be economically efficient for them to conserve other factors of production and consume more energy for a given level of output. Also, changes over time in an indicator may not closely relate to changes in individual sectoral energy intensities. For a more refined measurement, selected measures of energy efficiency or intensity from each sector or subsector are preferred.

Nevertheless, broad comparisons among countries can be made. Historically in developed countries, one of the driving forces for improving energy efficiency has been the relative increase in the real price of energy, compared with other economic factors. In the United Kingdom, the 42 percent increase in real GDP since 1970 has been accompanied by an almost static primary-energy demand. While part of this is related to the introduction of a cleaner alternative fuel—natural gas—and to some structural changes in industry, improved energy efficiency has played a major role. It has been estimated that since 1978, some 40 percent of the reduction in the U.K.'s energy intensity has been due to specific energy-efficiency initiatives and investments. Similarly in the U.S., partly as a result of energy-efficiency measures, there has been only an 8 percent increase in energy use since 1973, while GNP has increased 46 percent.

Box 6. High Power System Losses in Bangladesh

System losses have been excessive in Bangladesh throughout IDA's power lending program of six projects in the country over eleven years. Losses were 35 percent before 1979, 33 to 43 percent on an annual basis during 1980–1988, 46 percent (the peak) in October 1987, and 43 percent in May 1990. These high loss levels are continuing, despite a covenant in the fiscal 1988 Transmission and Distribution Project requiring loss reduction to 32 percent, and a second tranche release covenant in the fiscal 1989 Energy Sector Adjustment Credit that also required a loss reduction to 32 percent.

Due to high system losses and poor collections, a January 1991 supervision report states that payments for electricity reflect only 57 percent of the energy generated. Because of this lack of progress, IDA suspended lending for new energy projects in 1990 and suspended disbursements for ongoing projects in September 1991. he conclusion is that fundamental sectoral change is needed and without it power loss reduction patch work will not succeed in Bangladesh.

Source: World Bank reports.

of the possibilities of delaying capital-intensive investments in energy supply, and second, because of the potential savings in fuels. On average, energy supply and end-use efficiencies and industrial energy efficiency rates are two-thirds to one-half of what would be considered best practice in the developed world, regardless of the process involved. Without detailed country-specific analysis the exact magnitude of energy savings that can be achieved through supply- and demand-side efficiency and conservation measures is not known. However, *World Development Report 1992*, presents an energy-efficiency scenario in which a number of achievable economic and institutional reforms—in pricing and reduction of transmission and distribution losses and unused capacity—are assumed to be put in place to improve energy efficiency. Under this efficiency scenario, annual electric power investment requirements in the year 2030 would be about half what they would be under an unchanged-practices scenario, and emissions of pollutants would be about 35 percent less.

Several other technical studies have also estimated that at current relatively low energy prices and with the present state of technology, a savings of 20 to 25 percent of energy consumed by the existing capital stock in many developing countries could be achieved without sacrificing the economic benefits of energy use. This estimate reflects known opportunities that are attainable and cost effective. Over the longer term, as investments are made in new capital equipment, it is likely that larger energy savings will be achieved on the order of 30 to 60 percent above what is now possible with current equipment.

On the basis of 1990 energy consumption in the former USSR, Eastern Europe, and developing countries, a 25 percent energy efficiency saving would amount to saving 1 billion toe per year. At current oil prices, this would amount to $160 billion. The Bank's 1990 study on energy demand[3] estimated potential savings of primary energy from energy efficiency in eight developing countries (Brazil, China, India, Indonesia, Malaysia, Pakistan, the Philippines, and Thailand) to be about 100 million toe at current consumption levels, equivalent to $16 billion per year through the mid-1990s. This could be achieved by structural and energy price

Box 7. Syria: Energy Efficiency Improvements in Cement Production

The Syrian cement industry, the most energy-intensive industry in the country, recently accounted for about 34 percent of the energy consumption of Syria's industrial sector—this compares with 2 to 6 percent for the cement industry in most developing countries. The cement industry is government-owned and operated. A recent study of three plants managed by three of the seven government cement companies examined energy consumption patterns and constraints and identified areas where large improvements in energy efficiency could be achieved. The study concluded that Syrian cement plants were about 30 percent less efficient than similar plants in many other developing countries and identified the following categories for improvements:

- operational improvements involving no financial investment
- modifications requiring small investments
- long-term investments

It was estimated that operational improvements alone would result in fuel and electric energy savings of about 3 percent and 15 percent, respectively, on average and would increase production by 30 percent. For the three plants examined, this would translate into fuel-oil savings of about 8,800 tons and electricity savings of almost 80,000 megawatt hours. Undertaking the identified programs of investments would lead to almost equal savings of electricity and four times the savings of fuel oil. At the same time, local environmental pollution would be reduced as particulate emissions decreased.

Source: ESMAP reports.

Figure 7. Comparative Electricity Tariff Level in Current 1988 US$

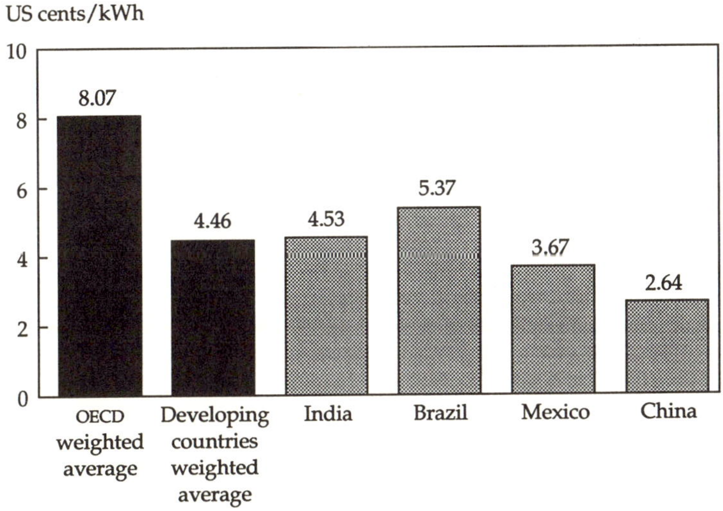

Source: World Bank data.

change, a faster rate of technology transfer, and fuel substitution. Savings were measured by evaluating the difference in the energy demand projections on the basis of a with or without energy-efficiency scenario. The difference between the two projections by 2010 represents a potential savings of some 750 Mtoe per year in the eight countries. At current oil prices, this would amount to $120 billion in savings (in 1992 dollars) per year from the late 1990s onward.

4

Why Is Energy Efficiency Performance in Developing Countries So Poor?

Bank experience in comparing developing- and developed-country performance in energy efficiency has highlighted four critical factors that directly correlate with differences in the efficiency of energy production and end use:

- the tendency to price energy, particularly electricity, below the costs of production and distribution, while developed countries mostly tend to recover the full costs and sometimes substantially more;
- the fact that many industrial and large commercial sectors in developing countries are dominated by a relatively few large monopoly or highly protected state enterprises, while developed countries tend to have less protected, more competitive industrial and commercial markets;
- the tendency of developing country governments to treat their monopoly public energy enterprises as a direct extension of government, with little differentiation among the ownership, regulatory, and management functions; while developed countries tend to have a more formal and transparent relationship where the rules of the game are more clearly spelled out; and
- the fact that barriers to the efficient functioning of markets are more pronounced in developing countries due to a relative lack of information intermediation, which results in relatively higher costs of information, financing, and management expertise.

Box 8. Summary of Energy Sector Characteristics in Developing versus Developed Countries

	Energy consumption	*Energy prices*	*Market structure for energy use*	*Energy supply-side institutions*	*Information barriers*
Developing countries	Low per capita consumption	Low	Protected industries	Public monopolies	Relative lack of intermediation of information, technology, and finances
	High growth rates	Subsidized	Public monopolies	Command and control regulation	
			Bias against efficiency in financing	Opaque accountability	
OECD countries	High per capita consumption	Market based	Competitive markets	Public and private enterprises	Market-based information, technology, and financial intermediation
	Low growth rates		Easy entry and exit	Transparent regulation	
				Checks and balances	

Box 9. Pricing of Residential Electricity

Even though electricity consumption in developing countries tends to be concentrated outside the residential sector, the impact of setting prices to reflect costs for residences can be large. A recent cross-national study of urban household energy use in twelve developing countries illustrates the importance of both income and price for residential electricity consumption. The study showed that all income groups are responsive to the price of electricity. For every 1 percent increase in electricity prices relative to kerosene, consumers will lower their consumption of electricity by about 0.9 percent in the poorest income groups and by 0.6 percent in the highest income groups.

The effects of income are just as strong as the effects of price. For every 1 percent increase in the higher-income groups there is about a 0.7 percent increase in consumption. The conclusion is that electricity pricing policies are also very important for residential consumers in developing countries, and they are relatively more important for the poor than for wealthy households when it comes to electricity consumption. Because the poor use electricity in very low quantities—mainly for lighting—pricing policies should probably have some kind of lifeline rate for lower-income groups.

Source: Barnes, D. 1992.

Poor Country Policy Environment

Low energy prices

Prices in most developed countries are set with the objective of covering at least the full financial costs of supply, but in many developing countries, energy prices (other than for some petroleum products) do not cover the economic or full financial costs of supply. This means that energy consumers—many of whom produce other goods—do not face prices that encourage them to use energy efficiently, select the most economic fuel, or use the technology that would best meet their needs.

In the transport sector, fuels are used inefficiently when fuel prices are below border prices and when there are distortion-producing price differentials among alternative fuels (gasoline, diesel, kerosene). In the case of electricity, a recent survey of electricity tariffs in sixty developing countries showed that on average, tariffs declined between 1979 and 1988 from 5.2 cents to 3.8 cents/kWh in constant 1986 U.S. dollars—or less than half the level of those in OECD countries—while power supply costs in those developing countries tended to be higher. The survey found that average tariffs for nearly 80 percent of the utilities in those countries did not cover the long-run marginal cost of supply. In addition to encouraging wastage in energy end use and making many energy-efficiency, conservation, and technology initiatives financially unattractive, these tariff policies strain the sector financially. This strain leads to inefficient, underfunded operations and maintenance practices, which in turn cause further energy wastage on the supply side.

Lack of competitive markets

Bank experience also indicates that the efficiency with which energy is consumed is directly related to the existence of private sector firms operating in competitive markets. This factor is particularly important in developing countries, where the household sector consumes relatively little energy compared to the commercial and industrial sectors. It has been well documented that many industrial processes in developing countries with protected industries require far more energy per unit of output than do processes that yield similar products in the industrialized world. For example, steel and ammonia fertilizer production often require twice as much energy per unit of output; pulp and paper production often requires three times as much. But in the absence of competitive markets, raising the energy prices paid by these protected industries will do little to encourage them to curb energy wastage or to adopt more energy-efficient processes and technologies.

Box 10. A Protected Industry

The Hadisolb Rehabilitation Project in Egypt failed because of *protection from competition*. Energy consumption, which under the project was expected to decrease by 37 percent, actually increased by 13 percent. Hadisolb faced no competition from either local or foreign sources because of government controls on steel prices, which were below the CIF price of imports. Hadisolb was also subsidized through low energy prices, low iron ore costs, and an exemption from paying royalties on its mining operations. The low levels of operating efficiency, as measured by labor productivity, product quality, energy efficiency, capacity utilization rates, and conversion rates, stemmed fundamentally from lack of basic management systems and procedures; and from laws and regulations that severely inhibited management initiative and enterprise, allowed no autonomy, and demanded little accountability.

Source: Gandhi *et al* 1993.

Box 11. Command and Control

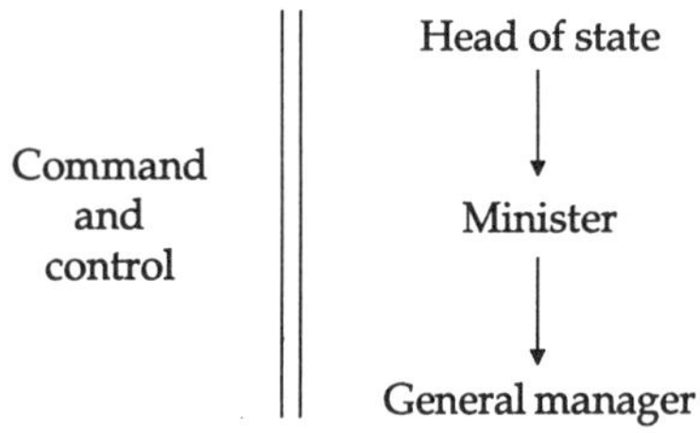

In a closed command and control environment it is politically difficult to:

- Make unpopular tariff decisions
- Resist the temptation to meddle in sector investment and management decisions, to use the utility as a vehicle for political patronage, to invest in *new* rather than maintain or renew *old*, to divert funds, to ignore bothersome environmental issues, etc.

A related point is that many of the protected industries in developing countries tend to be energy-intensive industries in the first place—cement, steel, refineries. In the absence of protection, it is possible that a much less energy-intensive mix of industries would have evolved.

Closed Command-and-Control Supply-Side Institutional Structures

When energy sector programs and projects appear technically sound but fail to deliver results, the reasons are, in many instances, attributable to weak institutions, absence of trained manpower, lack of an adequate legal framework, interventions by governments, uncertain and variable policy frameworks, and a command-and-control decisionmaking process that increases the likelihood of corruption and waste. Many needed energy sector reforms are never achieved if laws are not enforced or if there are severe delays in the settlement of claims. Sector reforms can also fail if accounting systems are so weak that budgetary policies cannot be implemented or monitored or if lax procurement encourages corruption, which in turn distorts energy sector investment priorities.

The Bank's new 1993 electric power policy paper[4] has concluded that while all of these constraints are significant, the most frequent cause of failure is the absence of a clear and transparent regulatory process. In a closed command-and-control system, many governments in developing countries are tempted to interfere in day-to-day organizational and operational matters that should be left to the control of enterprise managers. Such interference has tended to undermine the accountability of those responsible for management functions by influencing procurement decisions, mitigating against least-cost fuel choice, restricting the ability to raise power tariffs to meet revenue requirements, restricting the enterprise's access to foreign exchange, tying salaries to low civil service levels, and promoting excess staffing and political patronage.

In many cases, these problems have brought about generally inadequate energy enterprise management and organization, a lack of accountability, little concern for the environment, the flight of experienced and capable staff because of uncompetitive employment conditions, weak planning, inefficient operation and maintenance, high technical and nontechnical losses, poor collection, and weak financial monitoring and controls. Also under such circumstances, consumers, environmentalists, and other interest groups have little opportunity to be heard in setting policies regarding the investment program, pricing, access to service, utilization of service, energy conservation, plant location, and environmental issues.

Box 12. Electricity Pricing: Lessons Learned

The following is a quote from the December 1991 World Bank Project Completion Report for the Republic of Yemen Third Power Project.

"The Bank's flexibility in handling the financial issues as demonstrated by the downward revision of financial targets was not reciprocated by the Government and the Yemen General Electricity Corporation, who instead repeatedly cited social reasons for not taking decisions to raise tariffs. An important lesson from this project is that although relaxation of financial conditionality may be justifiable on occasion to allow continuation of policy dialogue, such relaxations should be made only if the Government shows its good faith by taking at least some tariff action, and when there is evidence of a strong and clear commitment on the part of Government to take further action to deal with the sector's financial problems. [In the future] it might be useful for the Bank to require upfront satisfactory progress in electricity tariff reforms prior to loan approval."

Barriers to Efficient Markets

The costs of adopting more energy-efficient systems, procedures, and technologies tend to be higher in developing countries because of a lack of intermediaries to reduce the costs of information, financing, and management assistance. This scarcity of intermediation is partly a result of the large gap between the higher cost of energy and the lower subsidized prices at which it is sold. There is very little incentive for profit-making enterprises, firms, or institutions to develop or to serve as market intermediaries to bring about a reduction in these costs.

Other factors affecting energy efficiency include inadequate legal structures through which intermediaries can work; the lack of or inability to enforce codes and standards; indiscriminate barriers to trade (tariffs and restrictions); large foreign exchange premiums; and discriminatory taxes on, or subsidies to, various energy forms, supplies, or substitutes, which distort market choices and cause wastage. Discriminatory subsidies can include domestic financing options. For example, an electric or gas utility that is able to borrow at a below-market rate (6–10 percent) to increase energy supplies will have an advantage over a competing energy service company, private power project, alternative fuels project, or firm that must pay the market rate (15–20 percent) to install more efficient motors or air conditioning. With regard to the trade sector, significant impediments to efficient production and end use of energy include trade restrictions or import duties on energy-efficient technology, equipment, and appliances and restricted access to foreign exchange.

5

What Can Be Done: Country Policy Priorities

Integrated Energy Strategies

Decisions on country priorities for taking action to improve energy efficiency and conservation in developing countries should be made within an overall integrated energy strategy. The formulation of an integrated energy strategy is the responsibility of a country's government. Governments must set policies and formulate strategies that point the energy sector toward the most efficient, equitable, and environmentally-compatible resource use that is feasible. Upstream decisions on both the energy supply and demand side must be made about sector structure, institutions, ownership, financing, fuel availability (coal, oil, gas), technology availability (import restrictions), structures of end use markets, and so on. Strategies currently in use or being discussed range from, at one extreme, supporting subsidized command-and-control public monopoly energy suppliers that sell to protected state enterprises that in turn engage in cost-plus pricing to final consumers, to, at the other extreme, a combination of public and private power suppliers wheeling power over common carrier transmission lines to competitive final consumers and local distribution companies.

In formulating integrated energy strategies, governments must first review both supply- and demand-side options and set priorities that address major impediments to improving the efficiency with which energy is produced and consumed. It is clear from the discussion in chapter 4, that the highest priority for improving the efficiency of energy supply and end use in many developing countries must be to improve the basic institutional and efficiency-incentive structures relating to energy. In many developing countries the formation of an integrated energy strategy would first result in targeting demand-side issues, such

Box 13. Deregulation and Import Restrictions

Some of the objectives of the Industrial Energy Efficiency and Modernization Project in India, initiated in 1990, were to develop and implement open-market policies and programs that would secure improved incentives for efficient energy use; build stronger institutions in the public and private sector to carry out energy conservation programs; carry out demonstration projects in key subsectors; and conduct a program for research and development needs.

The Indian government proposed to promote this plan by:

- deregulating the domestic economy to increase competition and provide stronger incentives for industry to use the various factors of production, including energy, more efficiently;
- adjusting the cost-plus pricing formulas for most industries to allow enterprises to retain financial savings derived from investments in energy conservation;
- reducing tariffs on energy-saving devices to lower the capital cost of energy efficiency investments;
- subsidizing plant-level energy audits and feasibility studies;
- subsidizing R&D programs and expenditures by both public sector research institutes and the private sector;
- adopting fiscal incentives for energy-saving devices and systems, including 100 percent depreciation allowance for income tax purposes, excise tax exemptions, and customs duty exemptions for solar and wind energy generation equipment; and
- de-licensing equipment manufacture for minihydro and microhydro systems, electric vehicles, solar, wind, and other renewable-source power-generating equipment.

Source: Gandhi *et al* 1993.

as pricing energy, promoting competitive markets in which energy consumers must operate, and taking selected information and technology-related interventions to help markets work more efficiently. On the supply side, the highest priority would be to make energy-supply enterprises responsive through institutional and regulatory reform and increased private sector participation, as well as to target plant rehabilitations, reduce transmission and distribution losses, and so on. Other priority areas, such as facilitating the ease of technology transfer and increasing the focus on energy efficiency in transport, would involve initiatives on both the supply and the demand sides.

In addressing these priorities, governments must think long term and maintain consistent policies so that their strategic sector objectives will ultimately be met.

Countrywide Incentive Structures

Energy prices: the first priority

In a developing country context, the highest priority structural initiative must be to let energy prices reflect the real costs of supply, including pollution abatement and damage costs. As part of the overall energy pricing regime, power suppliers and gas suppliers should pursue active load-management programs that are responsive to time of use, type of user, and quality of supply. While the impact of price changes may not be large in the short run, the oil price shocks of the 1970s and early 1980s show that the longer-term efficiency impacts can be quite dramatic in competitive markets.

The World Bank has extensive experience in encouraging developing countries to let energy prices reflect the real costs of energy supply. In fact, few issues between the Bank and its borrowers are as contentious as pricing, and the Bank's efforts to have prices reflect international levels or cover long-run marginal cost (LRMC), by imposing covenants

Box 14. Institutional Barriers

The bulk of the energy sector reform program of the Third Structural Adjustment Loan in Côte d'Ivoire was not implemented, mainly because the fundamental institutional problem in the sector—unclear lines of authority among different ministries and agencies—was never resolved. Consequently, no progress was made in the proposed formulation and implementation of institutional measures to strengthen planning capabilities and improve coordination in the sector. Likewise, no other activities were undertaken that required a cooperative institutional structure for their success; in particular, investment programs were not developed, and price and tariff policies were not elaborated. Two years later, however, the government did tackle the country's energy-institutional issues through the Energy Sector Adjustment Program and instituted wide-reaching energy reforms leading to a substantial improvement in energy efficiency.

Source: Gandhi *et al* 1993.

that set out financial targets, have often been unsuccessful. In the electric power sector, tariff levels are in principle targeted to reflect, on average, the power system's LRMC. However, in most instances the tariffs do not reach or remain at that level very long, and even if they do, distortions caused by cross subsidies result in actual prices that do not reflect costs. Consequently, many developing-country power utilities continue to be economically and financially nonviable and continue to require financial transfers from the national budget. Mobilizing additional resources for investment, facilitating a competitive environment, improving efficiency, and developing economically and financially-viable power sectors all require that electricity pricing be moved towards commercial practices.

Many utilities in developing countries have yet to market electricity actively or to do so with an energy efficiency focus. Energy enterprises should be encouraged to adopt marketing strategies to improve plant load factors, reduce peak loads, and reduce the need for high-cost peaking plant in order to minimize the construction of new power plants and to avoid burning fuel needlessly. All of these involve vigorous demand-side management (DSM) programs. Bank experience in proposing explicit DSM programs is limited; a review of past Bank power sector lending found that of 135 projects that contain explicit energy efficiency components in addition to promoting prices that reflect long-run marginal costs, only about twenty contained an explicit DSM tariff structure program. It is clear that the Bank must do more to encourage its borrowers to engage in aggressive market-oriented pricing strategies.

Making consumers responsive through competitive markets: another high priority

To stimulate supply and end-use efficiency, developing countries must also reduce barriers to the development of dynamic private commercial and industrial sectors operating in competitive markets. Protective industrial and commercial barriers must be eliminated, and trade restrictions and foreign exchange controls should be phased out.

Numerous studies suggest that new, more energy-efficient technologies are introduced more rapidly into countries in which competitive market forces are the rule. These studies have shown that countries in which the private sector is allowed to operate, subject to the discipline exerted by competition, are usually the first to adopt new technologies effectively. It is also clear that in the absence of competitive markets or other performance-based incentive structures, the availability of state-of-the-art technology alone will not have a great impact on energy efficiency. In many developing countries, current technical knowledge and modern technology are available, and some multilateral and bilateral aid programs have supported the technology-transfer process. However, such technology is frequently used inefficiently.

A recent study undertaken by the World Bank and several bilateral agencies investigated the transfer of a relatively simple energy technology, the diesel generator. This study reviewed the experience of a number of developing countries with operating diesel plants. In all cases it was found that when technology is available, the most critical variables determining success or failure are quality of staff and management, degree of accountability, degree of autonomy, and effectiveness of performance incentives. In one country, for example, identical diesel sets operated by a public utility and by a private company produced widely differing performance.

Studies on a macro level show that countries that have developed by competing internationally (the newly industrialized countries) have done so mostly on the basis of importing readily available technologies. Numerous country studies at the firm level have also shown that gains in productivity cannot come from technology alone unless appropriate institutional and competitive market-incentive structures are in place to identify and manage the use of a technology.

The Bank has had considerable experience in recent years in working with countries to open up markets and encourage competitive forces. Typical structural adjustment loans (SALs) and some sector adjustment loans (SECALs) over the past few years have contained such objectives as improving efficiency of resource allocation, strengthening export competitiveness, reducing or abolishing controls on imports, rationalizing currency exchange rates, and increasing the emphasis on price incentives and market forces.

Making Supply-Side Institutions Responsive Through Institutional and Regulatory Reform[5]

The creation of institutional frameworks to encourage efficient energy production and distribution must include initiatives on at least two levels: restructuring energy supply enterprises, and facilitating a transparent regulatory mechanism between government and the energy-supply enterprises.

The energy sectors in many developing countries are structured with a single national electric or gas utility or oil company operating as a public monopoly. This model is based on the principle that energy is a strategic and publicly-provided good and that people have a right to energy at low prices. This public monopoly model was suited to energy sectors in their formative stages because it facilitated expansion of

Box 15. Korea: Focus on Evaluating Management

The major power producer in Korea is KEPCO, a parastatal company responsible for most generation and all transmission and distribution. At the time of the first World Bank loan in 1979, KEPCO was considered a relatively well-managed company, but it was subject to excessive government regulation and control. It also was experiencing high staff turnover because of inadequate salaries and benefits. Coordination of its many operating units was posing problems, and concern was growing that its original organizational structure, established in 1961, was no longer appropriate to maintain or increase efficiency. It was also noted during the Bank's appraisal of the first loan that KEPCO needed to improve its planning and organization, that the pricing structure was inadequate, and that there were no clearly defined financial goals.

A subsequent Bank-financed consultants' study concluded that government control over KEPCO was excessive and that KEPCO should be allowed greater autonomy for managing its affairs. As a result, the utility made internal organizational changes. In 1984 the Government-Invested Enterprise Management Act introduced a new management structure for KEPCO and other public enterprises. The act gave a new external regulatory board the authority to separate policymaking from executive functions. KEPCO's president and management became fully accountable to this external regulatory board.

The act also provided the board with criteria for evaluating KEPCO's management performance and for awarding salary increases based on performance. The board operates under the scrutiny of the Public Enterprise Management Council, which reviews all public enterprise performance. The act allows substantial incentives for superior performance and stresses evaluating management, not company, performance.

The effects of this new regulatory structure have been striking. Between 1983 and 1988, KEPCO's operating costs declined in real terms by some 34 percent. This reduction was achieved in spite of rapidly rising real-wage costs and after taking account of changes in the conversion efficiency of new generating plant and of reductions in fuel costs during that period. In addition, technical measures on the supply side, and demand-management policies, such as mandatory time-of-use tariffs, are estimated to have reduced peak-load demands between 1977 and 1987 by about 800 MW, or some 7 percent of actual 1987 peak demand. Another characteristic of the new regulatory policies was that over time, tariffs were regularly adjusted to cover actual costs, including a substantial portion of capital investment costs.

Source: World Bank reports.

energy supplies, captured technical economies of scale, and, at least in the early years, made effective use of scarce managerial and technical skills. In some industrial countries (for example, France), where a performance-based regulatory system is firmly in place and where the utility borrows heavily in private capital markets and is therefore subject to capital market discipline, the model still functions satisfactorily.

In the context of many developing countries, however, as time passed, this structure resulted in a mixing of the roles of the government as operator, regulator, and owner of energy enterprises and drew governments into day-to-day management and unwarranted intervention in enterprise operations. It is also of increasing concern that such a command-and-control approach does not provide the opportunity for energy consumers, investors, or the environmental community to articulate their concerns. One alternative structure would involve setting up a more independent and transparent regulatory body to assist in redefining the roles of government, energy enterprises, consumers, and other interest groups. This implies a shift away from opaque command-and-control type regulation and toward decentralization and market-based incentives. Government would, of course, retain responsibility for setting objectives and articulating overall policies and would also establish the legislative and legal framework to protect the interests of the various stakeholders and the public. With a more transparent regulatory structure, consumers, investors, environmentalists, and other interested parties could all have a voice in determining policies relating to investment programs, pricing, access to and reliability of service, energy conservation, plant location, petroleum procurement, refining, distribution, and environmental issues.

Essential features of such a regulatory framework are transparency and openness; clear articulation of the reform objectives, including tariff policy; and a legal structure that clearly defines the procedures for reducing government involvement in management and increasing the autonomy and accountability of energy enterprise directors and managers. Such a framework also defines entry and exit conditions for investors and competitive enterprises.

In addition to the general principles of transparency and independence from day-to-day government interventions, regulation should address environmental issues. It is fortunate that the technology for addressing one of the most serious pollution problems of electric power production emissions of particulate matter is relatively simple and inexpensive. Increased use of gas-fired power stations will also be important in this respect. Where coal is the preferred fuel, constructing tall chimneys, siting power stations away from large population centers, and using emissions control devices all help to increase amenities and reduce hazards to health. Given the costs to life and health of particulate matter emissions and the modest costs of reducing these emissions to low levels, the case for working toward high standards of abatement is unambiguous. Developing effective regulatory institutions to address these issues will take time, but concerns about potential impediments and delays should be no excuse for inaction.

Developing-country energy-sector strategies must also promote commercialization, corporatization, and inceased private-sector participation. For energy enterprises to operate on commercial principles, they must be treated the same as commercial enterprises in the private sector. They should pay interest and taxes; earn commercially competitive rates of return on equity capital; and have responsibility for their own budgets, borrowing, procurement, salaries, and conditions pertaining to staff.

One way to broaden the financial base of energy-sector enterprises would be to switch some portion of their borrowing to financial intermediaries. Given the potential importance of the energy sector for capital market development, and the comparative price and income stability that will come through proper regulation and commercialization, the energy sector has the potential, through financial-intermediary lending, to transform domestic savings into investments in long-term bonds and equity issues of energy companies. This is one of the primary means by which capital markets in the United States and Western Europe have developed.

In some of the less developed countries with weak public and private sectors, undeveloped capital markets, and a relative lack of market forces, an early step in bringing about energy sector reform and increasing sector management efficiency would be to import international services into the sector under management contracts, twinning arrangements, or local concessions. Potential areas for contracting out services include plant maintenance, billing, revenue collection, vehicle maintenance, pipeline laying, line stringing, and pole and tower fabrication. Other areas for outside involvement include reducing supply-side system losses and increasing plant availability. In fact, in the petroleum sector in Sub-Saharan Africa, it has been estimated that liberalizing procurement, eliminating government monopolies, and instituting more transparent pricing could save the region up to $1.3 billion per year. The financial costs of these initiatives would generally be low relative to their economic benefits to the countries and their populations.

6

What Can Be Done: Sector Priorities

Making Markets Work Better

Even if better institutional and regulatory frameworks are in place, energy prices reflect real costs, and competitive end use markets begin to function, experience in both developed and developing countries has shown that other market imperfections can still create significant barriers to efficient energy production and end use. These include:

- Lack of a government track record on consistency, predictability, and credibility in policies to encourage energy efficiency.
- Information gaps on energy losses, loss reduction techniques, technology and process options, financing, and joint venture opportunities, all due to a relative lack of market intermediation and resulting high transactions costs. In the household sector in particular, energy users do not usually have easy, low-cost access to necessary technical information and capital, and first-time appliance buyers do not generally have the sophistication to understand the potential differences in the costs of ongoing energy consumption.
- Household end-use energy consumers usually not facing real costs of energy use because households are often not adequately metered and because investment decisions are often split among tenants, owners, and contractors.

In addition to these barriers, other factors tend to be associated with lower levels of energy efficiency. These include:

- The general availability of energy-inefficient appliances, equipment, and structures, due partly to an absence of minimum energy-efficiency codes and standards for commercial buildings and for small consumer items such as appliances and motor vehicles; and a weak institutional capacity to enforce such codes and standards.

- The fact that, for a variety of reasons, end-use energy consumers tend to have higher implied discount rates than do energy suppliers. There is evidence to suggest that commercial enterprises and households tend to have much higher discount rates than do supply-side utilities and that household sector discount rates are inversely correlated with income. This has led some countries to introduce specific payments for new types of energy-saving investments by electricity customers, paid by either the government or, as in several U.S. states, the electric utilities themselves.

In developing countries, industrial advisory services have sometimes identified ways of reducing energy consumption per unit of output, as well as other costs. Such initiatives are important for improving energy efficiency, but their success, too, will depend greatly on prices that reflect the full economic and environmental costs of energy. These will, in themselves, help make energy-efficient technologies financially more attractive to industry and individuals.

Where competition exists, market barriers tend to be less marked in the commercial sector and are lowest in the industrial sector because these sectors operate in an environment where awareness of costs and benefits is important. Indeed, in developed countries with competitive markets, the industrial sector and especially energy-intensive industries have substantially improved their energy efficiency as economic growth has encouraged rapid stock turnover and the introduction of new, more efficient technologies.

Nevertheless, a recent International Energy Agency study showed that while energy-efficient equipment for industry in OECD countries is less subject to market barriers than equipment for other end uses, a number of significant barriers still slow the penetration of energy-efficient equipment into the energy sector. The most serious barrier is the lack of information about availability and reliability of new equipment. Other important barriers include a separation of those who bear the costs of new equipment from those who benefit from it, limited capital, rapid payback requirements dictated by investment opportunities elsewhere, the impact of electric and gas tariffs, lack of interest in peripheral operating costs, and legal and administrative obstacles. Private transport suffers from many of the same barriers as the household sector. The fuel economy of a transport vehicle is usually not the most important criterion in a purchase decision.

In developing countries, options for addressing barriers such as these include:

- Increased demand-side management;

- Setting up or strengthening national, local, or industrial energy-resource centers or efficiency institutions to help improve training and information intermediation and promote demand-side management initiatives;
- Putting in place energy-efficient standards and codes;
- Targeting specific technology transfer and efficient fuel-use options; and
- Focusing more on energy efficiency in transport.

Demand-Side Management

Demand-side management in industrialized countries

Every electric utility is subject to demographic patterns in energy usage unique to its service area. The utility can graph this usage, or load. *Load factor* is a key measurement that compares a utility's average kilowatt-hour load to its peak, or maximum hour's usage, in a given year. Raising the load factor increases the utilization of plant capacity, and this can be brought about by reducing the peak load on the system. DSM strives to achieve this effect by changing the patterns of electricity usage to reduce peak load or slow its growth and thereby defer the need for additional capacity to meet peak load.

Box 16. Replacing Power Plants With Glass and Plastic

In the ASEAN countries, over 30 percent of electricity is used by commercial buildings. In Bangkok, about half the electricity consumed in commercial buildings is for air conditioning. For a centrally air-conditioned office building, one square meter of low emissivity window would reduce the building's heat gain sufficiently to save $5 worth of electricity annually. The reduced air-conditioning needs would permit the use of smaller chillers and associated units, saving more than the initial cost of the special windows and resulting in a *negative cost to conserve energy*.

For $10 million, a low-emissivity window -oating plant can be constructed with an annual production of 2 million square meters of windows. Over their thirty-year lifespan, these windows would save 4 million TWhe, equivalent to the annual sales of an 800-MWe power plant costing more than $1 billion. For about $1 million a plant can be set up to manufacture windows locally, using rolls of imported coated plastic film.

Source: Lawrence Berkeley Laboratory.

The peak load can be changed by the utility or energy-supply enterprise through aggressive electricity pricing, with rates designed to discourage electricity usage at times of system peak. This pricing strategy would include seasonal rates, time-of-use rates, interruptible service rates, and a variety of other options depending on supply and demand conditions.

DSM can also be used to reduce the growth in overall energy consumption and thus defer the need for additional capacity to meet total demand. This is achieved by working directly with energy consumers and with energy-using equipment manufacturers to adopt processes, and adopt or produce equipment, that will result in more efficient consumption of energy. Such options include better building design and insulation; use of more energy-efficient light bulbs, motors, controls, window coverings, and appliances; manufacture and better maintainance of more energy-efficient transport vehicles, and so on.

While the first set of initiatives involving electricity pricing must be carried out by the electricity supplier (and the regulator when there is one), there are different models for encouraging the adoption of the second set of options. In fact, a variety of different private- and public-sector institutional formats have been developed that promote demand-side energy efficiency and conservation, sometimes in conjunction with the energy supplier but in many cases independently. All of these approaches require feedback, from demand-side changes to supply -ide planning and vice versa. In the U.S., this link between the supply and demand sides has, in fact, been formalized into a utility-based energy-service planning process known as Integrated Energy Resource Planning (IERP—see box 17). Both the IERP model and the various outside energy-efficiency institution models of many other industrial countries have had significant successes. No small part of the reason for these successes is the fact that in many industrialized countries, energy prices on average come close to reflecting the financial costs of energy (although not always the environmental costs), and the end-use commercial and industrial markets in which energy is consumed are generally competitive.

Demand-side management in developing countries

DSM is not currently pursued with much intensity in most developing countries (Singapore and, more recently, Pakistan are exceptions). The reasons vary from country to country but generally revolve around the facts that:

- energy prices are low and subsidized;
- end-use markets are not highly competitive;

Box 17. Integrating Demand-Side Management With Supply-Side Planning: The U.S. Experience

Historically, U.S. electric power utilities mostly met demand growth by adding generating capacity. Loads were served as the need arose, with little effort to manage demand. That approach to resource planning stood in contrast to the practice of industrialized Asian and European countries with comparable growth rates in electricity demand. Constraints on power supplies, in particular limited access to fuels, led many industrial countries to view increases in loads less as needs to be served than as incentives for shaping loads to fit available resources.

Beginning in the early 1980s, however, pressures from regulatory commissions, together with the high financial and environmental costs of building additional capacity, caused the concept of demand-side management to begin gaining acceptance in the U.S. Its evolution is part of a planning process called Integrated Energy Resource Planning. IERP is primarily a U.S. process whereby utilities and their regulatory commissions work jointly to evaluate available demand- and supply-side options (including purchased power) and determine an optimal energy service strategy, given economic and environmental factors. IERP puts all possible sources for meeting load growth on a "level playing field." Conservation and other efficiency investments are acquired in a planning process that uses the same discout rate for these investments as is used for making supply-side investments. In fact, the essential concept of IERP is the "equal" treatment, or integration, of energy-based and conservation-based energy service. Planners attempt to rank by cost all of the different energy supply and end-use technologies, processes, and programs that might be used to provide energy services, and implement them, beginning with the lowest-cost opportunities.

The concept is appealing. IERP is a tool that has been used in the U.S. to promote energy-efficiency investments as an alternative to capacity expansion. In the U.S., this is generally performed by the utility. IERP, as practiced in the U.S., has usually developed in an environment in which most end users purchase their electricity in a noncompetitive market and where a strong regulatory regime that can allocate costs and benefits across consumer groups is in place. For example, if consumers are charged a price of 8 cents per kWh but the peak cost to the utility is 10 cents, the utility—and presumably the country—can benefit by discouraging demand growth and use during the peak period.

Regulatory commissions encourage the utility to negotiate with customers on ways to reduce their overall consumption. The benefits and costs from this reduction are distributed in various ways among the utility, the customers, rate payers in general, and the taxpayer. Usually

(Box continues on the following page.)

Box 17 (*continued*)

the risks and costs are mostly borne by the rate payers because regulators have allowed the utility to write off the costs of these demand-reducing programs against the general rate base. In many of these programs, utilities use their preferred and sometimes subsidized access to capital markets to finance equipment and services to eligible customers. There are a variety of permutations of this approach, but they all require the utility to identify investments or practices by customers that would reduce the electricity consumption of consumers and thus avoid new additions to capacity.

In the United States, implementation of IERP activities is forecast to reduce electricity demand by about 45,000 MW of generating capacity in 2010 and 90,000 MW in 2030. The net economic benefit would be about $35 billion for the 1990–2030 period.

- the energy-supply enterprises are weak institutions that have major difficulties even in supplying energy and collecting bills;
- regulatory bodies do not exist; and
- there is a lack of knowledge about and high-level support for DSM initiatives on the part of government.

These barriers must all be addressed if DSM is to have a significant impact on developing countries' energy consumption.

In pursuing DSM initiatives, it is clear that, as in the industrial countries, developing-country energy suppliers are the ones who must pursue the energy pricing options. In fact, with the strengthening of institutions, most developing countries, having few explicit or legally-constrained regulatory policies, can more easily pursue aggressive load-management programs designed to alter the shape of the prevailing load curve by reducing peak demand and encouraging a more economical demand mix. Such programs would include pricing to reflect differences in time-of-use costs, classes of customer, location, and types of loads. Pursuit of better DSM pricing policies could yield large gains in developing countries even where average prices do not yet cover full costs.

With regard to the nonpricing DSM options that require working closely with energy consumers and equipment manufacturers, the few developing countries that have seriously pursued at least some of the available options have largely attempted to create some form of semi-autonomous energy-efficiency institution not directly associated with the energy-supply enterprise. How far the more ambitious integrated planning process, as it has developed in the U.S., can be applied effectively in developing countries is yet to be determined. In many developing countries, in addition to a lack of regulatory agencies to police utilities, low energy prices and poor reliability of service often weaken the incentives for consumers to respond to DSM initiatives. In concept, a large part of the potential gains from demand-side management can come from suppliers' ability to work with some of their larger commercial and industrial customers to agree on complementary actions that benefit both parties. In many instances, these actions can be specified in contractual arrangements. However, in many developing countries, the services of the energy suppliers are so poor that the larger customers

Box 18. Residential Demand Side Management for Thailand

A demand-side management (DSM) assessment was recently completed for the residential sector in Thailand. The investigation encompassed a comprehensive analysis of the major residential end uses of electricity: space cooling, refrigeration, lighting, cooking, water heating, and the powering of other appliances. The study outlined twenty-three economically-viable DSM measures that could be achieved through the use of existing technologies. If fully implemented, these improvements could reduce annual electricity use by up to 500 GWh and coincident peak electrical demand by 160 MWe during the first year of a residential DSM program.

The highest potential savings could come from improving the energy efficiency of refrigerators. Improvements to the insulation and compressors of Thai refrigerators were predicted to reduce electrical use from 400 kWh to below 200 kWh per unit per year. Possible countrywide savings could be up to 170 GWh per year if more efficient refrigerators were introduced into the marketplace. Negotiations are underway with a large Thai refrigerator manufacturer to produce a high-efficiency prototype unit for testing and evaluation.

Estimates for residential housing savings were based on replacing less efficient equipment in existing housing as it is retired from service and by instituting cost-effective energy-efficiency measures in new construction. With full implementation over ten years as new buildings are constructed and inefficient equipment in existing buildings is replaced, such a strategy could result in cumulative savings of more than 6,000 GWh and peak reductions of more than 2,000 MW by 2005. For the sake of comparison, the overall electrical consumption of the residential sector in 1989 totaled 7,025 GWh, while Thai households were responsible for approximately 20 percent of the peak utility load of 7,095 MW.

Source: Florida Solar Energy Center.

either provide for their own needs or have substantial standby capacity for meeting the frequent system failures. In the absence of improvements in reducing the frequency of outages, few of these customers would be willing to rely on contractual arrangements with an electric power utility.

Also, few energy-supply enterprises in the developing world have the necessary customer information to begin aggressive non-pricing DSM programs. Most of them have little or no end-use data, and many do not even have data on customer use by class. They are also not strong enough institutionally to undertake such managerially- and administratively-intensive programs. Many power utilities are in serious financial difficulty; they have trouble simply collecting their own bills and operating and maintaining their systems. As a first step, these utilities will require substantial institutional reforms, which are now a matter of high priority in many countries and are the focus of World Bank support. At the same time, most utilities will also need increased private sector involvement and concentration on major plant rehabilitations and transmission and distribution loss reduction programs. They will also need to begin building a DSM end-use customer database. In the meantime, in many countries alternative institutional arrangements will have to be pursued to encourage the implementation of non-pricing DSM options. Some of these alternative institutional arrangements are discussed in the next section.

As DSM programs are increasingly put into place, lessons will begin to emerge with respect to the potential benefits to developing countries of integrating supply- and demand-side measures. With respect to implementing end-use efficiency, creative solutions are needed to overcome market and institutional barriers to conservation. World Bank lending in the energy sector should be based on and, where necessary, support (as part of country assistance strategies) the development of integrated energy strategies. These would help borrowing countries take advantage of all energy-supply options, including cost effective conservation-based supplies and renewable energy sources.

Energy Efficiency Institutions

In OECD countries, even where energy prices tend to reflect costs and end-use markets are competitive, it has been demonstrated that information programs help narrow the gap between the technical potential for energy efficiency and current efficiency levels by providing consumers with the technical, economic, and financial information they need to make decisions on energy consumption. Experience in several developing (Korea, Tunisia, Pakistan) and developed (Japan, U.K., the Nether-

Box 19. Conservation Center in Pakistan

The energy conservation center, ENERCON, strengthened under the Pakistani Energy Sector Loan, provides a good example of how institutions can contribute to energy efficiency. ENERCON completed energy audits in forty-three private-sector industrial plants, where potential average savings are estimated at about 22 percent of the total energy consumption, with a payback period of about six months. ENERCON also completed audits in twenty-seven public-sector industrial plants. A building energy code was drafted. Forty-three preliminary audits of small buildings and a detailed audit of a large government office complex were completed. There were audits of 300 agricultural tubewells and retrofitting of 100 of them. Also, extensive and successful energy conservation training and outreach programs were conducted, including seminars and workshops for about 1,800 participants, and a power plant was rehabilitated. Currently, efforts are underway to commercialize the highest energy-savings projects identified by ENERCON.

Source: Gandhi *et al* 1993.

lands) countries has shown that an effective way to address some of the information intermediation and policy barriers to efficiency is to set up an independent, high-level energy-efficiency institute or resource center to intermediate the information market. Particularly in developing countries—where most energy-supply enterprises resist taking responsibility for end-use energy efficiency and conservation because conservation conflicts with their objectives of increasing sales and revenues—such centers or institutes are in some cases best located apart from the major energy-supply enterprises. If given enough stature, they can begin to provide checks and balances to the supply-side forces.

Specifically, independent, high-level energy resource center(s) or efficiency institution(s) can serve as the institutional focal point for conservation, efficiency, and alternative fuel initiatives in a country. Depending on their composition and focus, they could carry out some combination of the following demand-side information dissemination, technical assistance, and technical and financial intermediation functions:

Information dissemination

- Disseminate information on technology options, financing, and successes and failures;

- Promote and assist with demonstration projects, such as introducing more efficient light bulbs, motors, building design, window coverings, and solar and wind power;
- Provide contacts for energy audits; and
- Provide training, information, and advice on loss-reduction techniques.

Technical assistance

- Where energy prices reflect costs, help establish private energy service companies that share in the profits from loss-reduction or efficiency-increase initiatives;
- Carry out energy audits when private sector auditors are not available;
- Help government or the regulatory authorities draft codes and standards for buildings, refrigerators, air conditioners;
- Identify and lobby against macroeconomic and sectoral barriers to successful conservation and alternative fuel initiatives (import duties or restrictions on more efficient technologies, foreign exchange controls, protected industries that engage in cost-plus pricing, etc.); and
- Serve as a focal point for drawing on energy-efficiency technical assistance from bilateral aid agencies, NGOs, and programs such as ESMAP.

Technology intermediation

- Promote technology intermediation: identify technology needs or opportunities in the country and put outside business firms that use or offer up-to-date, efficient technologies in touch with companies that need technology assistance; and
- Provide an intermediation function for energy-service companies.

Financial intermediation

- Receive, appraise, and bundle labor-intensive, low-capital-requirement, efficiency, conservation, and alternative-fuels projects for potential World Bank, commercial bank, and other donor funding.

Bank-financed projects have included support for the establishment of energy conservation centers, the strengthening of existing institutions to promote conservation, and technical assistance for energy efficiency surveys and energy audits. The Bank has supported energy conservation

Box 20. Eastern European Efficiency Centers

Four energy efficiency centers are now in operation to encourage conservation in the emerging democracies of Poland, the Czech and Slovak Republics (formerly Czechoslovakia), and Russia, which rank among the least energy-efficient countries in the world. The Czechoslovak Center for Energy Efficiency (Prague) was founded in December 1990, the Polish Foundation for Energy Efficiency (Warsaw and Katowice) was created in January 1991, and the Russian Center for Energy Efficiency (Moscow) was recently registered in January 1992.

The efficiency centers were founded to help these countries achieve economic development and environmental protection by promoting energy efficiency. The centers are designed to tap regional expertise and draw on international experience to advise high-level energy policy makers and implement energy-efficiency measures on a broad scale. Each center is a non profit, independent institution run by recognized local experts in the area of energy conservation. They employ scientists, economists, and engineers to encourage energy efficiency through:

- policy analysis and support
- joint-venture development
- training and demonstration
- public education services

During the first year of operation, the Polish and Czechoslovak centers: facilitated adoption of least-cost utility planning as an official government policy (Poland); drafted an appliance energy labeling system; developed an efficient lighting project (Poland); produced a television series on energy-saving measures in households (Poland); prepared popular and technical articles on energy conservation for "Universe" and "Green Slovakia" (Czechoslovakia); initiated an energy-efficiency technologies database (Czechoslovakia); and presented policy advice to parliament (Czechoslovakia).

Priority activities for 1992 in all three countries include:

- investment analysis and partnership identification to create links between local and Western firms interested in cooperating in the area of energy efficiency;
- analysis of potential energy-saving technical projects, such as producing energy-efficient compact fluorescent lamps, and evaluating efficiency of district heating systems;
- establishing a database on technology and energy-efficiency measures to facilitate joint initiatives; and
- energy conservation legislation to advise policymakers on energy-resource pricing, conservation potential, and mechanisms for implementing energy-saving measures.

Source: Battelle, Pacific Northwest Laboratories.

institutions in a number of countries, including Argentina, Bangladesh, Barbados, Brazil, Cyprus, Hungary, India, Indonesia, Korea, Pakistan, Portugal, and Senegal. These institutions were created or supported through a variety of instruments, including SALs, energy sector loans, various types of energy projects, industrial energy conservation projects, and refinery projects.

The success of World Bank efforts to promote energy efficiency through dedicated efficiency institutions is related to a country's energy prices, the existence of competitive markets, government attitude, credibility, and consistency of policies, and the scope of the efficiency institutions mandate. In countries where governments have sought World Bank assistance in developing or strengthening broad-scope national institutions and have given them high priority and visibility, such as in Pakistan (ENERCON) and Korea (KEMCO), projects have been successful. In cases where government did not follow through on support for institutional development or where the energy conservation center was a relatively minor player in a larger government unit, there were varying degrees of failure. For example, while a primary objective of the Cyprus government was to wean the country from its near total dependence on imported fuel with the aid of an energy planning and conservation project, the government gave the organization in charge of energy planning and conservation a relatively minor role in decisionmaking and then abandoned the reform efforts when oil prices started to drop.

Private-sector energy-service companies have been slow to take off in developing countries. While the concept of hiring such a company to undertake an energy audit of a business firm, implement the highest-return energy-saving measures, and share in the resulting energy savings is appealing, several barriers exist. Not least of these are subsidized energy prices and the inadequacy of the legal and contracting framework in many countries. Another is the shortage of trained technical personnel to staff energy-service companies.

In a few of the higher-income, more advanced developing countries, possibilities may also exist for instituting some form of integrated energy-resource planning that institutionalizes the review of demand-side options in routine energy-supply planning. Experience suggests, however, that in many countries this will be a difficult task. Many of these energy-supply enterprises have major problems managing their supply business without undertaking administratively-intensive consumer-support programs. In general in a developing country context, it may sometimes be better to channel this type of support through a dedicated energy-efficiency institution operating in conjunction with both the energy-supply companies and the private sector.

Box 21. KEMCO: A Multidimensional Energy Efficiency Institution

The Korea Energy Management Corporation (KEMCO) was established in 1980 by the National Assembly of the Republic of Korea to encourage energy-efficient development of the national economy. KEMCO employs more than 600 staff, and its budget has quadrupled in ten years, from $5 million in 1981 to $20 million in 1990. The corporation is entirely government-owned but receives only 30 percent of its budget from the national government. The remainder is provided by revenues from its services to residential, commercial, and industrial consumers.

KEMCO has a wide range of programs underway in all of the major end-use sectors. Its ten-year experience is unique among developing countries, permitting it to pursue advanced energy-efficiency concepts, from research to implementation and evaluation. KEMCO's current portfolio includes:

- energy audits and technical assistance programs;
- financial assistance, through the administration of dedicated monies from the National Energy Rationalization Fund (generated from taxes on imported energy);
- promotion of co-generation and district heating projects;
- support of energy-conservation R&D activities;
- provision of general education, training, and information services;
- inspection and approval of energy-using equipment and materials; and
- promotion and utilization of new and renewable energy sources

To date, KEMCO's industrial, co-generation, and district heating programs are estimated to have produced economic savings of more than $3.5 billion. KEMCO offers a valuable model to newly-industrializing Asian countries that want to establish an energy-efficiency institution.

Source: Byrne *et al* 1991.

Standards and Codes

Another important aspect of a country's regulatory framework is minimum appliance and vehicle standards and building codes. In OECD countries, regulations and standards have been widely applied to appliance manufacturing and to thermal standards in the buildings sector and more selectively by imposing an up-front capacity charge on building designers equal to the additional capacity cost that the building will impose on the utility systems. However, there is still scope for further regulatory action to promote energy efficiency. For instance, vehicle fuel

Box 22. Improved Stoves: Problems, Benefits, and Solutions

The modern, efficient biomass stove is an important step toward self-sufficiency and a higher standard of living for the millions of people who have access to low-cost, readily available biomass fuels. Fuel savings from improved stoves can reduce cash outlays, diminish walking time to collect fuel, reduce air pollution released into the environment, and alleviate local pressure on wood resources.

In light of the benefits of improved stoves, the question can be asked why adoption rates have been so poor. Early programs assumed that if improved stoves were presented to people, they would quickly be adopted and the intervention would lead to self-sustaining programs. There are several reasons why this often did not happen. One was the uncritical belief that efficiencies obtained in laboratories would translate into efficiency in the home. Also, some programs introduced stoves into regions where people did not buy either the stove or the fuel, an obvious failure of identifying the market for the stove. Price was also a significant barrier to adoption, especially in areas where there was little cash for stoves or fuel. Donors in many cases gave short-term funds rather than the sustained, longer-term funding that the programs needed.

An example of the benefits of improved stoves is the program in urban Rwanda, where stove adoption has been high. Three years after the project started, about 25 percent of households in the capital were using improved stoves. One reason for the success was that the charcoal price in Rwanda was relatively high. As a consequence, the potential savings from an improved stove were dramatic; the stove can pay for itself in fuel savings in less than one month. Also, the stoves could be purchased in every market outlet and in many general stores.

The success in Rwanda provides some lessons. First, the programs that are successful have focused on groups of users that would most likely benefit from improved stoves, including those paying significant amounts of money for fuel and those who must walk long distances to collect fuel. Subsidies for the stoves themselves do not seem particularly effective since they create an artificial demand for the stoves, which may not actually be used by the household. However, external support for such activities as testing laboratories or consumer surveys can be successful. Programs are more effective if there is significant interaction between those who design the stoves, those who produce the stoves, and those who are going to use the stoves.

Source: Barnes *et al* 1992.

efficiency standards have so far been applied or set out as recommended norms only in a few countries—Australia, Canada, Jordan, the U.K., and the U.S.

Where regulation is necessary, it should, wherever possible, employ economic incentives to achieve its goals rather than attempt to legislate behavior without changing the underlying structure of private incentives. Nevertheless, in developing countries where information costs are high and numerous other market barriers exist, promoting end-use efficiency among the large number of geographically-dispersed commercial buildings and small-scale consumer groups through building codes, upstream standards for manufacturing, and importing such appliances as refrigerators and air conditioners, may be the preferred approach. Such programs usually have a relatively small investment component but require a large and sustained technical assistance effort over the long term. In most cases, public institutions will have powers to enforce codes and standards, but considerable scope exists for expanding the participation of consumers and the general public.

If standards are to be set, laboratories will be required to measure standards; producers, consumers, and governments will have to agree on acceptable standards; and the institutional mechanisms will have to be strong enough to support the enforcement of those standards. Lack of agreement and weak enforcement mechanisms have been serious impediments to the effectiveness of past efforts to improve the efficiency of buildings, electric motors, and consumer appliances.

Technology Transfer and More Efficient Fuel Use

These two components comprise a fourth set of initiatives to improve directly the efficiency with which energy is produced and consumed and to reduce the financial, economic, and environmental costs of energy production and consumption.

With regard to improving energy efficiency and reducing financial and economic costs, there is the technical potential for large efficiency gains using technologies that are readily available. For end uses such as refrigeration and lighting in the household and commercial sectors, improvements of 30 to 70 percent are technically possible. There is a somewhat lower potential for space conditioning (indoor-climate control) and water heating, although improvements in the shell of new buildings show substantial promise. Technical changes could contribute to significant improvements in the fuel economy of road transport vehicles, on the order of 15 percent, although achieving greater efficiency, on the order of 30 to 50 percent, would require significant modifications in vehicle attributes. Technological improvements are also possible in a variety of industrial processes and for a range of cross-cutting industrial technologies, although their potential varies case-by-case. For instance, in developed countries it has been estimated

Box 23. Transport Energy Savings

The Chilean Urban Streets and Transport Project focused more on energy conservation than has any other Bank transportation project. It included construction and evaluation of a pilot segregated bus/tramway in Santiago to demonstrate cheaper alternatives to future expansion of the metro system; construction and evaluation of a pilot bikeway; and traffic-management measures, such as bus priority lanes in Santiago, Concepción, and Valparaiso to reduce congestion and pollution.

The project was designed to reduce deterioration of the urban street infrastructure, improve the cost effectiveness of maintenance, and greatly reduce vehicle operating costs by implementing pavement/maintenance management systems throughout the country. Of course, while the Chilean Urban Streets and Transport Project provides great expectations, it is too early to tell whether these will actually lead to tangible results, given that actions have to be taken by various municipalities that were not party to the original agreements with the Bank.

Source: Gandhi *et al* 1993.

that currently available improved industrial motor systems can save 30 percent over current levels. The potential savings in developing countries would be far higher.

The reasons technology transfer and economically-justified fuel switching do not occur more rapidly are the same reasons discussed in this paper for overall poor energy efficiency on both the supply and the demand side. There is, however, another facet to the energy technology transfer issue in developing countries—how to reduce pollution. As pointed out in chapter 3, improved efficiency in energy production and consumption will help, but the decisive measures are abatement technologies and the policies that bring these technologies into use. The significant policies include using such environmental taxes and regulations as an incentive for the energy industry and its consumers to adopt cleaner fuels (such as gas) and clean fuel technologies (such as particulate emission controls, and, where merited, scrubbers or advanced combustion technologies).

Technological advances have put developing countries in a better position to reduce all forms of pollution from electric power generation than the industrial countries were in as recently as twenty years ago. In industrial countries the capital stock takes about thirty years to turn over, and retrofitting is costly. Because developing countries are making new investments, they have the opportunity to install less polluting plants right away. Employing economic and institutional reforms to

Box 24. Efficient Resource Allocation

If an objective of World Bank policy is to improve the allocation of resources in developing countries—including nonpriced environmental resources—Bank policy should ensure that energy conservation is efficient. An energy-conservation investment is economically efficient if the cost of reducing the demand for a unit of energy is no greater than the cost of supplying that amount of energy. If one accepts the objective of improving the allocation of resources, the desirable level of investment in energy conservation must therefore be that amount at which the cost of avoiding the need for a unit of energy is just equal to the cost of supplying that unit—where *cost* is taken to mean the *full social costs,* including the environmental costs.

It is important to emphasize that the economically-efficient level of investment will be less than the level that is technically achievable. Not all energy savings that can be achieved are actually worth achieving. This means that the Bank must also be careful about setting physical targets for energy efficiency. It is difficult to estimate the magnitude of economically-efficient energy savings, for that will depend on how consumers and businesses respond to improved signals; and behavioral responses of this kind are difficult to predict. Governments must first concentrate on making certain that the signals are right rather than on attaining a particular arbitrary level of efficiency.

improve the efficiency with which fossil fuels are used and progressively adopting environmentally-beneficial technologies not only reduce local pollution appreciably but also improve economic efficiency.

But using cleaner fossil fuels, more energy-efficient technologies, and improving overall energy efficiency will not by themselves solve the long-term problem of stabilizing or reducing carbon dioxide accumulations in the atmosphere. That would require a much greater use of nuclear or renewable energy. While the relative costs of nuclear power have increased, developments in renewable energy in the 1970s and 1980s—in solar, wind, and biomass energy, in particular—have led to remarkable cost reductions in these technologies. There is now a growing awareness that renewable energy is an abundant resource that increasingly can be harnessed.

Greater Focus on Energy Efficiency in Transport

Another area of potentially large energy savings is the rapidly growing transport sector. The annual oil consumption in developing countries is estimated to be 833 mtoe, of which about half is used in vehicles. Due to its dependence on liquid fuels, the transport sector has now become a significant burden on the balance of payments of most oil-importing countries. Fifty percent, or $74 billion, of the energy bills for these countries in 1988 was for transport.

Road transport is particularly sensitive to changes in the oil market situation, since it is one of the key sectors that is likely to be dependent on petroleum products for many years to come. Thus, efforts to achieve efficient use of oil-based fuels in road transport are of continuing importance in most developing countries. This issue is even more serious in non-oil-producing developing countries, where road traffic uses a greater share of total imported oil than it does in developed countries. On average, road transport in developing countries consumes 82 percent of total transport energy. Likewise, the share of energy consumed by the transport sector is usually greater in developing than in developed countries. Transport accounts for 15 percent of total energy consumption in the European Community, 26 percent in the U.S., and between 30 percent (Kenya) and 50 percent (Sri Lanka) in developing countries, many of which still use a combination of traditional and modern transport technologies. Within the road subsector, cars consume about two-thirds of the energy in the developed countries and much less in developing countries. In the latter, buses and trucks frequently account for more than 70 percent of energy consumed on the road, as opposed to 20 to 25 percent in developed countries.

In both developed and developing countries that rely heavily on non-motorized vehicles, transport energy use is a smaller share of total energy use. In South and East Asia, where nonmotorized vehicles are in substantial use, transport accounts for 24 percent of energy use. In China, the transport share of total energy use is only 7 percent, in part due to long-term policies favoring energy-efficient nonmotorized and rail transport. The EC enjoys more energy-efficient transport than the U.S., due in part to its generally greater reliance on walking, cycling, and public transport. Africa's high transport energy share of 40 percent can be attributed, in part, to the absence of intermediate and nonmotorized vehicles as options to fill the niche between walking and relying on buses.

In the cities of developing countries, motorized vehicles are a significant source of airborne toxic pollutants, accounting for up to 95 percent of lead emissions. Three factors make pollution from vehicles more serious than in industrial countries. First, many vehicles are in poor condition, and lower-quality fuels are used. Second, motor vehicles are concentrated in a few large cities. In Mexico and Thailand about half the vehicle fleet operates in the capital city, and in Brazil a quarter of the fleet

Box 25. Supply-Side Reforms

The Bank policy paper on electric power, *The Bank's Role in the Electric Power Sector: Policies for Effective Institutional, Regulatory and Financial Reform* (1993), outlines an approach that can be more broadly applied to much of the energy sector. The guiding principles for the Bank set out in that paper are:

- A requirement for all power sector loans will be explicit country movement toward the establishment of a legal framework and regulatory processes satisfactory to the Bank. To this end, in conjunction with other economywide initiatives, the Bank will require countries to set up transparent regulatory processes that are clearly independent of power suppliers and that avoid government interference in day-to-day power company operations (whether private or publicly owned). The regulatory framework should establish a sound basis for open discussion on power-sector economic, financial, environmental, and service policies.
- In some of the least developed countries, the Bank will assist in financing importation of power services to improve efficiency.
- The Bank will aggressively pursue the commercialization and corporatization of, and private sector participation in, developing-country power sectors.
- The Bank will focus lending for electric power on countries with a clear commitment to improving sector performance in line with the above recommendations.
- To encourage private investment in the power sector, the Bank will use some of its financial resources to support programs that will facilitate the involvement of private investors.

operates in São Paulo. Third, a far larger percentage of the population moves and lives in the open air and is thus more exposed to automotive pollutants.

The poor are usually the most affected. They and their children are more likely to walk than to ride, and they are thus exposed to noxious fumes and to lead, which is known to affect mental development and the neurological system. Lead and other pollutants also contaminate food in open-air restaurants, which are frequented by the poor. The lead problem is being tackled effectively and relatively cheaply in some countries; concentrations have gone down 85 percent in the U.S. and 50 percent in Europe over the past two decades. *World Development Report 1992* describes how lead emissions from vehicles in developing countries could rise fivefold over the coming few decades—or could fall to negligible levels. Policy choices account for the difference.

With respect to the energy efficiency of motor vehicles, this will not only require standards (often a blunt, command-and-control imperative) but also greater efficiency in taxation, including the possibility of taxes on vehicle fuels. In Europe and Japan, taxes on gasoline, for example, range from $3 to $4 per gallon, while in the U.S. and in many developing countries, taxes are relatively low (in some countries the fuel is subsidized). European taxes on diesel fuels are lower, but still greater than in the U.S. and most developing countries. As a result, it is not surprising to find that the average fuel efficiency of vehicle fleets in Europe and Japan is up to 50 percent higher than in the rest of the world.

Until the mid-1980s, the objective of improving road capacity was frequently addressed in Bank transport projects through traffic management, without mention of fuel efficiency—although the former usually leads to the latter. Approaches to traffic management include segregation of motorized and nonmotorized traffic, encouragement of the wider use of bicycles and development of special facilities for them, creation of vehicle-free precincts for pedestrians, incentives for greater investment in and use of public transport, incentives for higher vehicle-occupancy rates, and parking controls. Schemes of this kind can reduce vehicle fuel consumption in metropolitan areas by more than 30 percent. Cities in China, Ghana, Indonesia, Japan, and the Netherlands are all considering such schemes, with a greater emphasis on nonmotorized traffic and pedestrian facilities.

Traffic can also be restricted through quantity-based measures, such as the area traffic bans based on license plate numbers introduced in Athens, Mexico City, and Santiago. These, however, are only stopgap measures and can sometimes make the situation worse, since the people with higher incomes simply purchase a second vehicle, and a market for fake license plates develops. A third possibility is some form of congestion pricing, such as area licensing, access fees to city centers, higher fees and taxes on parking during business hours, and electronic road pricing. Despite the very successful example of the Bank-financed Singapore Area Licensing Scheme, *congestion pricing* thus far has been more discussed than implemented.

Any measure that supports the faster movement of traffic, particularly public transport, can result, at least in the short run, in energy savings. This is indeed a focus of the Bank's urban transport activities. In contrast to the U.S., where less than 10 percent of all trips are on mass transport, in the developing world this figure in many instances exceeds 80 percent, with buses on public streets being the most common transport mode. Improving the energy efficiency of bus transport rests partly on the efficient operation and maintenance of the buses but, more important, on the traffic they must face on the roads.

Finally, World Bank projects have shown that in the transport sector, particularly with regard to roads, fuels will tend to be used inefficiently if one or more of the following conditions applies: fuel prices are below border prices; there are distortion-causing price differentials among substitutable fuels (gasoline, diesel, kerosene); vehicles or roads are poorly maintained; there are a large number of old vehicles; or traffic is poorly managed (avoidable congestion, wide variations in speed). To address these issues more explicitly, Bank transport project appraisals could routinely compare all fuel prices with border prices; examine price differentials and consider steps to prevent fuel mixing or the substitution of diesel cars for gasoline cars; describe efforts, such as vehicle inspection, to improve vehicle and road network maintenance; describe efforts to discourage the purchase of high fuel-consumption vehicles through differential taxes or license fees; report steps taken to liberalize the importation of more efficient engines or vehicles; and describe efforts at traffic management, including implementing bus lanes, congestion charging, and perhaps bicycle paths.

Bank urban projects have also shown that one of the problems in improving efficiency in the transport sector is that city layouts are not planned, nor do people select where they live and work solely on the basis of minimizing transport or energy costs. In fact, to date there is not a consensus about how to approach urban planning in developing countries or how to solve the transport problem. In any case, the tools available to influence the outcome of the urbanization process are limited, and their application is subject to political, institutional, and economic uncertainties.

The Bank is committed to assisting countries in dealing with problems of urban areas and transport in particular. The issues and approaches advocated by the Bank are well set out in *Urban Transport: A World Bank Policy Study (1986), and in Urban Policy and Economic Development: An Agenda for the 1990s* (1991). Before new or different country policy and strategy prescriptions can be made, more analytical work needs to be undertaken on energy efficiency issues in transport.

7

Strategy for the World Bank

There is now a congruence of several forces in the developing world that makes timely the formulation of a strategy to address energy efficiency and conservation issues better. Today many countries are becoming more receptive to reforming the way energy is produced and consumed as they experience converging pressures from (see chapters 3, 4, 5):

- rapidly growing demand for energy
- major constraints on available energy financing
- increased pressures to sustain the environment
- poor energy-sector performance and unsatisfied customers
- reappraisal of the roles of governments and the public and private sectors in development

Because these factors are forcing developing countries to address long-neglected issues of energy wastage in production and end use, the Bank now has an opportunity to focus further on improving developing-country energy efficiency and conservation policies and practices.

The Bank will continue its efforts toward increased lending for components to improve energy efficiency and promote economically justified fuel switching (see chapter 2). In addition, however, in taking advantage of the increased receptivity of many developing countries to efficiency issues, the Bank will sharpen its focus by undertaking the following four-point program:

(i) To gain greater country commitment, the Bank will better integrate energy efficiency issues into its country policy dialogue so that they can be addressed at an earlier stage.

In the Bank's general country policy dialogue with developing countries, greater emphasis will be given to energy pricing and to fundamental institutional and structural factors that affect supply- and demand-side energy efficiency. The Bank will assist borrowing countries to develop integrated energy strategies that give consideration to both supply- and demand-side measures. These integrated strategies will also consider unconventional renewable options, incorporate environmental considerations, and include a long-term capital mobilization plan. The energy sector is a candidate for greater attention because of its size, its strategic role in the growth process, and its major environmental impacts.

The Bank will give greater prominence to energy-pricing issues in the general country dialogue and will continue to encourage developing countries to move vigorously toward a system of private firms operating in competitive markets in their commercial, industrial, and household sectors. The Bank will continue to encourage governments to eliminate protectionist policies, divest themselves of monopoly enterprises, eliminate barriers to free trade, reduce foreign exchange premiums, and move toward eliminating discriminatory energy taxes and subsidies. It is recognized that there are many competing claims for attention in the country policy dialogue and that not every issue can be addressed at the same time. In many cases, however, the adjustment process has reached the point where a greater focus on these issues is warranted. Energy efficiency is a high priority area where, at the policy level, the Bank has a clear comparative advantage to act.

(ii) The Bank will be more selective in lending to energy supply enterprises.

Governments should clearly demonstrate that they are putting in place structural incentives that will lead to more efficient energy supply and consumption. The Bank will not continue to finance energy-supply projects where poorly performing and highly polluting public energy enterprises and their governments are unwilling to carry out fundamental structural reforms that could significantly improve the ways they do business.

Reducing pollution from electric power production requires both improvements in efficiency and investment in abatement. The Bank will draw on its extensive experience in working with energy-supply enterprises to assist governments in developing an institutional structure that can, by promoting efficiency and compliance with environmental standards, respond effectively to changes taking place in the macroeconomic environment. The rapid growth of energy demand, even under the most optimistic of efficiency scenarios, will call for substantial new additions to capacity. In contrast to the developed countries, where growth rates are much lower, these additions in developing countries will form a larger and larger part of the supply stock. It is particularly important to ensure the efficient operation of this new, as well as the old, capacity.

Developing a more responsive institutional structure will not be an easy or quick task. Significant changes will be required in the way governments do business in the energy sector. A new system of governance needs to be developed and applied in the regulation and management of energy enterprises. The thrust will be to encourage the development of energy-supply enterprises that will respond to pricing and competitive pressures in ways that will improve the environmental and overall energy efficiency not only of the enterprises but of the economy as well. Increased private-sector participation will be encouraged. At the enterprise level, it is important to define clearly the responsibility for energy efficiency and to convince management that energy efficiency is a profitable use of management time and investment resources. Where local expertise is not available to restructure and set up regulatory mechanisms, the Bank can provide funding for technical assistance to shepherd movement in the desired direction.

(iii) Approaches for addressing demand-side management and end-use energy intermediation issues will be identified, supported, and given high-level, in-country visibility.

The Bank will increase its efforts to improve intermediation in the energy and industry information markets in developing countries to reduce the relatively high information, management, technology, and financing transactions costs. There is a role here for both the public and private sectors. As the gap between the cost of energy and the price at which it is sold is reduced or eliminated, market intermediaries will increasingly be able to earn a profit through arbitration of information, technology, financing, and management assistance. The Bank will play a role by identifying, supporting, and financing both public- and private-sector initiatives that can serve the intermediation function and pursue DSM objectives.

In some countries with more efficient monopoly power utilities, the utility-based Integrated Energy Resource Plan model might be pursued. In other countries, the initial model could be some form of utility subsidiary or the dedicated energy-efficiency institution structure outlined in chapter 6, or both. The public- and private-sector institutional framework and home(s) for demand-side end-use energy initiatives should be determined during project appraisal. A variety of different public and private-sector options are available, depending on existing country conditions. The point is that along with addressing the fundamental energy pricing, end-use market competitiveness, and supply-side restructuring issues, demand-side management and market information, process, and technology-intermediation functions must also be addressed within some form of specified institutional frameworks in the country.

If a form of semi-autonomous dedicated energy-efficiency institution is chosen, the Bank can assist with the initial terms of reference and the financing of such efficiency institutions or resource centers and with bringing international expertise to the institutions for twinning and for training of staff. Ultimately the Bank can lend funds to the institutions to support commercially viable energy savings and fuel-switching activities or programs that are approved and bundled by the institution(s). Other possibilities are to support information intermediation of commercial energy efficiency or fuel-switching initiatives through industry or trade associations or through some form of existing financial-intermediary institutional structure.

A key to the success of any energy efficiency institution is that after it becomes established and accepted, it should have an objective of at least breaking even financially. Whether this objective is achieved through government payments for specific services, fees, profit sharing, or attracting nongovernment industry or utility payments does not matter. The point is that financial discipline will help focus such institutions on initiatives that yield the highest returns and will force the institutions to work closely with energy suppliers, manufacturers, and private-sector energy consumers. However, public-sector institutions should not directly compete with the private sector in areas where the private sector can and does provide energy services effectively.

Finally, it is worth noting that developing countries do not need large capital investments to improve intermediation in the energy and industry information markets. Intermediation efforts, most of which are targeted at improving end-use energy efficiency, primarily involve policy and institutional initiatives relating to the information dissemination, technical assistance, and financial and technical intermediation functions noted in chapter 6. These initiatives are highly labor intensive, require relatively little in the way of capital, and should be carried out with major participation by developing-country personnel and the private sector.

Substantial amounts of capital might be required, however, for efforts to upgrade and retool industries and to establish new industries that produce more energy-efficient appliances, motors, controls, and so on. Presumably, most of this technology will be mobilized as a result of the information and technology intermediation process and will be financed by the private sector through direct investment or joint ventures with industrial-country firms.

Box 26. Hypothetical Energy Efficiency Institutional Structure in a Developing Country Energy Sector

Government/ regulator

- Create enabling environment
- Competitive end-use markets
- Eliminate/reduce import and joint venture restrictions
- Supply and demand side transparent rules of the game
- Supply side restructuring
- Codes and standards

Energy utility/supply companies

- Load management tariff structure
- DSM incentive programs
- Information programs
  processes
  technology
  audit information

- IERP
- Liaise with manufacturers

- Advocate and promote energy efficiency
- Identify and address barriers
- Intermediate information technology financing

Semi-autonomous efficiency institution(s)

(iv) The Bank will give greater attention to the transfer of more energy-efficient and pollution-reducing technologies in its sector and project work.

For all sectors, including basic materials-processing industries, the Bank will actively monitor, review, and disseminate the experience of new efficiency-enhancing supply-side and end-use products, technologies, and processes, and cleaner and pollution-abating technologies as they are developed and reach the marketplace; help finance their appli-

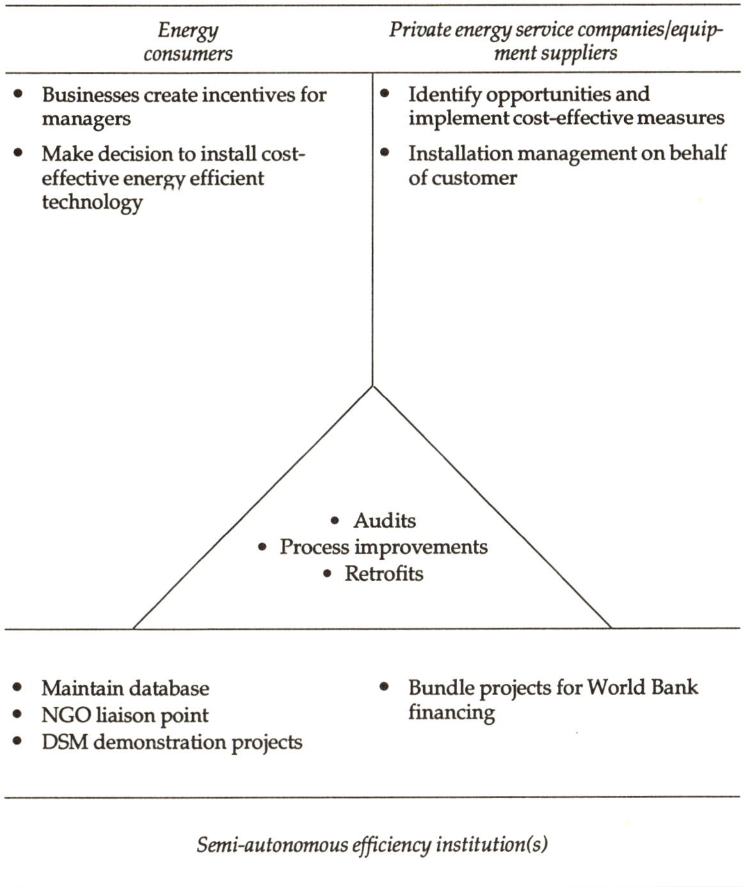

Note: The two triangles contain functions that could be shared by or allocated to any of the three institutions bordering on them.

cation; and encourage the reduction of barriers to their adoption. Staff working in all sectors will explicitly review technology options during project appraisals and in sector work. Cost effectiveness in pollution abatement is a required goal of policy, and the best mix of energy-efficient and pollution-abatement technologies will need to be found.

Historically, technical advances have led to large reductions in the amounts of energy required for any given purpose. But because costs also declined commensurately and because they were associated with far-reaching innovations, they were a source of expansion for the energy industry, not of contraction. Innovations in other fields and the growth of incomes and economic activity led to an almost endlessly increasing array of applications of commercial energy, compounding the effects of cost reductions via efficiency improvements on the industry's expansion. Examples can be found in all fields and in every decade of the past three centuries.

As outlined in *World Development Report 1992*, policies to mitigate the effect on the environment of energy production and consumption take two complementary approaches. The first uses economic instruments and institutional reforms to encourage the more efficient use of energy. Following from this, the second approach is either to develop technologies that reduce the polluting effects of conventional fuels or to use less polluting substitutes. The costs of pollution from industry, energy, and transport are already high and will grow exponentially if these problems are neglected. Encouraging energy conservation is a helpful first step in tackling pollution, but it cannot solve the problem alone. The effects of rising populations and incomes will soon swamp any reductions in demand per person. It is thus absolutely essential to reduce emissions per unit of production. This requires investment in new equipment and the development of new technologies.

Many developing countries, in fact, have a disproportionately high level of older, inefficient, and more polluting capital stock in all sectors. Partly because of low growth in the 1980s, turnover rates of capital stock have been low. This, together with the forecast of rapid growth in energy demand from relatively low levels, means that newly-installed supply- and demand-side equipment will be significant additions to existing capital stocks. Each new investment offers an opportunity to incorporate cost-effective pollution control. In ten years' time, new plants will account for more than half of the industrial output of developing countries and in twenty years for practically all of it. Thus, policies that lead to the adoption of a proper combination of low-waste processes and end-of-pipe controls should permit developing countries to reduce emissions from large industrial plants (as output expands) at a lower cost than is being incurred by industrial countries.

It is well known that available technologies in most sectors and for most purposes show widely different energy use levels for the same energy service, while not showing large differences in overall cost. As a result, there is the need to put in place policies, legislation, mechanisms, systems, institutions, and incentives that facilitate technology transfer and encourage the use of the most efficient competitive technologies. It can be argued that the actions discussed in this paper to bring the efficiency with which energy is produced and supplied in developing countries up to OECD standards constitute a difficult and lengthy process and that they will have only a one-time impact. Greater long-term potential for major improvements in the conversion of energy into environmentally-benign economic output lies in incentive structures or processes that channel new investment into the most up-to-date and efficient competitive technologies.

The most significant actions to bring about technology transfer in a developing country context have already been outlined in this paper. Nevertheless, other initiatives such as joint ventures, increased private sector participation, additional information dissemination, and putting in place energy-efficiency building codes, appliance manufacturing standards, and enforcement mechanisms, can be encouraged. Institutional barriers to the development of natural gas with low-cost, high-efficiency combined-cycle technology should also be targeted. Existing and emerging pollution abatement technologies should be identified, and policies, regulations, and taxes put in place to bring about agreed abatement levels. Over the next decade, many new energy-efficient technologies and processes, which are only now emerging, will become commercial. It is important that their transfer to developing countries take place in a timely manner.

While not directly part of an energy conservation or efficiency focus, the process of promoting energy efficiency should also include identifying opportunities for the economic development of alternative fuels such as mini-hydro, photovoltaics, and wind energy. Such alternative fuels options should be publicized and made available to the private sector.

The Bank has much experience with alternative energy, having financed numerous smaller alternative energy components in larger projects. The ESMAP program has also been involved in considerable renewable energy work, and more recently the FINESSE program (Financing of Energy Services for Small-Scale Energy Users) has been institutionalized in the Bank as the Alternative Energy Unit in the Asia Region. The Latin American and Caribbean Region is considering a similar action.

Endnotes

1. For purposes of simplifying the discussion, the terms *energy efficiency* and *energy conservation* are used interchangeably in this paper. Both are assumed to refer to the efficient production and use of energy, as in providing the same level of productive output with less energy and energy investment. They do not refer to sacrificing the benefits of energy services.

2. All dollar figures are U.S. dollars unless otherwise indicated.

3. Imran and Barnes, *Energy Demand in the Developing Countries: Prospects for the Future.*

4. World Bank Industry and Energy Department, *The World Bank's Role in the Electric Power Sector: Policies for Effective Institutional, Regulatory, and Financial Reform.*

5. This section draws heavily on *The World Bank's Role in the Electric Power Sector: Policies for Effective Institutional, Regulatory, and Financial Reform* (1993). This section may be skipped if the reader is familiar with that paper.

Appendix. Guidelines Used to Identify Bank Projects With Energy Efficiency Components

A preliminary review of some 1,500 World Bank Staff Appraisal Reports (SARs) was undertaken in an attempt to identify projects with energy efficiency components. The initial guideline used in screening projects for energy efficiency components was: does the component explicitly target energy efficiency, and will it result in a favorable change in output per unit of energy input. In reviewing projects it was clear that exactly what constitutes an energy efficiency improvement measure differs among the seven types of activities or areas reviewed: electric power, industry, transport, energy, structural adjustment, fossil fuels, and a category called "dedicated energy efficiency"—that is, those projects whose primary or sole objective was to address energy efficiency.

The issue of pricing deserves special mention since most World Bank projects address pricing issues in one way or another. For purposes of this review a very conservative definition of price-related energy efficiency objectives was chosen. Measures which raised prices, eliminated subsidies or reformed tariffs solely for the stated purpose of improving financial performance were not included. On the other hand, if pricing measures were designed to alter energy consumption behavior, they were included to the extent that they formed part of the projects that also directly addressed other energy efficiency objectives. Similarly, many projects include measures to improve institutional performance. Only those projects which specified improving energy efficiency as a reason for institutional change were included.

The following are examples, by lending instrument or sector, of project components that were included and excluded from the review.

Electric Power

Included were efforts to improve the efficiency of distribution and transmission, all retrofits to generators, process improvements, legislative and tariff changes when energy savings was an objective, conversion to other fuels and other efforts to cause consumers to use less energy.

Not included were extensions beyond existing networks and increases in capacity, even if more efficient generation replaced less efficient.

Industry

Included were measures which raised productivity without raising energy inputs, such as process improvements, legislative and tariff changes to promote competition, retrofits and conversions of plant machinery which did not entirely replace old machinery, and institutional upgrading toward an energy efficiency objective.

Not included were measures which tried to increase productivity or revive the sector without a specific objective of improving energy production or use.

Transport

While it can be argued that almost any measure in the transport sector can be used to enhance at least short term energy efficiency, from highway paving to vehicle maintenance to driver training, for purposes of this review only projects which stated improved energy efficiency as an objective were included.

Fossil Fuels

Included were projects which reduced spills, leaks, and gas flaring. Also included were efforts to improve the combustion of fuel.

Not included were measures which sought to raise fuel production solely to meet demand even if efficiency improvements resulted.

Dedicated Energy Efficiency Projects

This category was introduced to isolate projects which were initiated with, and designed around, the objective of improving energy efficiency as a means of achieving wider objectives.

Energy Sector and Structural Adjustment Loans

Included were reforms which were designed to promote the efficient production or use of energy, such as promoting competition, changing legislative and regulatory structures, and tariff reforms.

Not included were similar measures which were pursued for economic stabilization or other reasons exclusive of changing energy production or consumption patterns.

World Bank Activities in Support of Energy Efficiency and Conservation

	A *Direct interventions/ assistance*	B *Fuel choice and technology options*	C *Policy, pricing, and institutional issues*
Type of Bank activity	(i) Dedicated energy efficiency projects (ii) Projects with energy efficiency components (iii) Pre-investment and technical assistance activities; sector work	(i) Renewable/alternative energy projects (ii) Gas development projects (iii) Forestry projects	(i) Structural adjustment loans (ii) Sectoral adjustment loans (iii) Conditionality in project lending in power, energy, industry, and transport sectors (iv) Development and strengthening of institutions
Efficiency objectives addressed through:	(i) Direct interventions to conserve energy, e.g. through project lending that addresses: • plant rehabilitation • industrial efficiency • efficient technologies • retrofitting • transport efficiency • demand management • household energy efficiency (ii) Energy audits and technical assistance (e.g. through ESMAP and Bank project lending) (iii) Energy efficiency strategy development through sector work and ESMAP/IENED studies; assistance to governments in pursuing energy efficiency objectives	(i) Use of cleaner fuels and addressing rational fuel choices where feasible such as gas over coal (ii) Renewable energy development (iii) Fuel technology options that lead to efficiency and are environmentally benign such as combined cycle and cogeneration (iv) Sustainable fuelwood production through forestry and biomass projects	(i) Economy-wide reforms that remove inconsistencies and create incentive structures for energy efficient behavior sucha s through: • proper pricing • removal of trade barriers (ii) Regulatory reforms; minimum standards, standards and codes (iii) Development of institutional capability/strengthening of relevant ministries in countries, and the development of private sector to address efficiency issues more effectively

Bibliography

Africa Technical Industry and Energy Division, The World Bank. "The Importance of Traditional/Household Energy for the Energy Sector." *Cross Fertilization Brief* Feb. 1992.

Africa Technical Industry and Energy Division, The World Bank. "The Institutional Quagmire of the Power Sector in Africa." *Cross Fertilization Brief* Sept. 1991.

Africa Technical Industry and Energy Division, The World Bank. "A Regional Study on the Rationalization of Petroleum Supply in Sub-Saharan Africa." *Cross Fertilization Brief* Mar. 1992.

Ahuja, Dilip R. *Estimating Regional Athropogenic Emissions of Greenhouse Gases.* Washington, DC: U.S. Environmental Protection Agency, September 1990.

Anderson, Dennis. *The Energy Industry and Global Warming: New Roles of International Aid.* Nottingham, England: Russell Press Ltd./Overseas Development Institute, 1992.

Anderson, Dennis. *The Forestry Industry and the Greenhouse Effect.* A Report for the Scottish Forestry Trust and the Forestry Commission. Edinburgh: Scottish Forestry Trust, 1991.

Anderson, Dennis and Catherine D. Bird. "Carbon Emissions and Carbon Fixing from Economic Perspective." *Oxford Bulletin of Economics and Statistics* 54.1 (1992).

Andersson, Maria and Antoine Baya-Vuma. *Tortillas Must Be Baked on Flames: Stoves & Kitchens in Nicaragua—A Proposal.* Lund, Sweden: Lund Centre for Habitat Studies.

Aristodemou, Nicolas E. "Energy Planning and Conservation Project EC2 Brick Industry." Final Report. London: Ove Arup & Partners International Ltd., 1988.

Aristodemou, Nicolas E. "Energy Planning and Conservation Project EC4 Hotels, Hotel Apartments and Offices." Final Report. London: Ove Arup & Partners International Ltd., 1988.

Armstrong-Wright, Alan. *Urban Transit Systems: Guidelines for Examining Options.* World Bank Technical Paper No. 52. Washington, DC: The World Bank, May 1986.

Association for the Conservation of Energy. *Contracts for Energy Management: A New Approach to Energy Efficiency.* Paris: OECD, 1988.

Augenblick, Mark and B.S. Custer. *The Build, Operate, and Transfer Approach to Infrastructure Projects in Developing Countries.* Policy Research and External Affairs Infrastructure Working Papers. Washington: The World Bank, August 1990.

Bashmakov, I.A. and V.P. Chupyatov. *Energy Conservation: The Main Factor for Reducing Greenhouse Gas Emissions in the Former Soviet Union.* Richland, Washington: Pacific Northwest Laboratory, 1991.

Barnes, Douglas. *Understanding Fuelwood Prices in Developing Nations.* Forthcoming Industry and Energy Department Working Paper, Energy Series. Washington, DC: The World Bank, 1992.

Barnes, Douglas, Kevin Fitzgerald, and Gordon McGranahan. "Interfuel Substitution and Changes in the Way Households Use Energy: Estimating Changes in Cooking and Lighting Behaviour in Urban Java." *Pacific and Asian Journal of Energy,* 1:21-49. Also publishing in Industry and Energy Department Working Paper No. 29. Washington, DC: The World Bank, 1991.

Barnes, Douglas, Willem Floor, and Henry Peskin. *Accounting for Traditional Fuel Production: The Household Energy Sector and Its Implications for the Development Process.* Forthcoming Industry and Energy Department Working Paper. Washington, DC: The World Bank, 1992.

Barnes, Douglas, Keith Openshaw, Kirk Smith, and Robert van der Plas. *An Evaluation of Improved Biomass Cookstove Programs: Prospects for Success or Failure.* Forthcoming Industry and Energy Department Working Paper. Washington, DC: The World Bank, 1992.

Barnes, Douglas and Liu Qian. "Urban Interfuel Subsitution, Energy Use and Equity in Developing Countries." International Conference of the International Association for Energy Economics. Honolulu, 8–10 July 1991. Forthcoming in a special issue of *Energy.*

Bates, Robin. *Energy Conservation Policy, Energy Markets, and the Environment in Developing Countries.* Environment Working Paper No. 45. Washington, DC: The World Bank, May 1991.

Bates, Robin and Edwin Moore. "Commercial Energy Efficiency and the Environment." Background Paper No. 5, World Development Report 1992. Washington, DC: The World Bank, 1991.

Baum, Vladimir. "The Generation Game." *Petroleum Economist,* February 1992.

Beesley, M.E. and S.C. Littlechild. "The Regulation of Privatized Monopolies in the United Kingdom." RAND Journal of Economics 20 (Autumn 1989).

Behrens, Alfredo. "Brazil's Industrial Energy Conservation Programme." *Energy Policy* 13 (1985).

Behrens, Alfredo. "Household Energy Consumption in Rio de Janeiro Shanty Towns." Manuscript Report. Ottowa: International Development Research Centre, 1986.

Besant-Jones, John E. "Financing Needs and Issues for Power Sector Development in Developing Countries During the 1990s." International Conference of the International Association for Energy Economics. Caracas, Venezuela, June 1989.

Besant-Jones, John E. *The Future Role of Hydropower in Developing Countries.* Industry and Energy Department, Energy Paper No. 15. Washington, DC: The World Bank, April 1989.

Besant-Jones, John E. *Policies for Reforming Power Sectors in Developing Countries.* Washington, DC: Industry and Energy Department, The World Bank, forthcoming.

Besant-Jones, John E., ed. *Private Sector Participation in Power Through* BOOT Schemes. IEN Energy Series Paper No. 33. Washington, DC: The World Bank, December 1990.

Besant-Jones, John E., et al. *Review of Electricity Tariffs in Developing Countries During the 1980s.* IEN Energy Series Paper No. 32. Washington, DC: The World Bank, November 1990.

Besant-Jones, John E. and Lori Hylan, eds. *Managing Risks of Investments in Developing Countries.* IEN Energy Series Paper No. 55. Washington, DC: The World Bank, May 1992.

Bischoff, D. "Assessment of Future Potential Gains in Fuel Economy from Technology." Paper prepared for the IEA/OECD Rome *Workshop on Low Emission, Low Consumption Vehicles.* Washington, DC: National Highway Traffic Safety Administration, 1990.

Blair, Peter D., et al. *New Electric Power Technologies Problems and Prospects for the 1990s.* Washington, DC: Office of Technology Assessment, U.S. Congress, 1985.

Bleviss, D.L. *The New Oil Crisis and Fuel Economy Technologies: Preparing the Light Transportation Industry for the 1990s.* New York: Quorum Books, 1988.

Booth, Roger and Philip Elliott. "Sustainable Biomass Energy." *Shell Staff Selected Papers.* London: Shell Centre, London, 1990.

Boyle, Stewart, Linda Taylor, and Ian Brown. *Solving the Greenhouse Dilemma: A Strategy for the UK.* London: Association for the Conservation of Energy, June 1989.

Brand, M. "Impacts of the Structures of Electricity and Gas Tariffs on the Rational Use of Energy." *Energy Economics* 12.2 (1990).

Brinner, Roger E., et al. "Optimizing Tax Strategies to Reduce Greenhouse Gases Without Curtailing Growth." *Energy Journal* 12.4 (1991): 1-14.

Brookes, L. "The Greenhouse Effect: The Fallacies in the Energy Efficiency Solution." *Energy Policy* 12.2 (1990).

Brown, Lester R., et al. *State of the World 1991: A Worldwatch Institute Report on Progress Toward a Sustainable Society.* New York: W.W. Norton & Company, 1991.

Brown, Lorenzo, Michael Einhorn, and Ingo Vogelsang. "Toward Improved and Practical Incentive Regulation," *Journal of Regulatory Economics,* Vol. 3, p. 323–338, 1991.

Burns, Robert, Daniel Duann, Kenneth Rose, Kevin Kelly, and Narayan Rau. *Discussion Papers on Competitive Bidding and Transmission Access and Pricing Issues in the Context of Integrated Resource Planning,* NRRI, 1990. (Publication No. NRRI-90-9.)

Busch, J.F. "Saving Energy and Power in Thai Commercial Buildings." ASEAN Special Sessions of the ASHRAE Far East Conference on Air Conditioning in Hot Climates. Kuala Lumpur, Malaysia, 26–28 October 1989.

Businaro, L. and A. Fedrighini. "Prospects of Energy Conservation in Transportation." In *Energy Demand and Efficient Use,* edited by F. Amman and R. Wilson. New York: Plenum Press, 1981.

Butcher, David. *A Review of the Treatment of Environmental Aspects of Bank Energy Projects.* IEN Energy Series Paper No. 24. Washington, DC: The World Bank, March 1990.

Butera, Federico M. *Renewable Energy Sources in Developing Countries: Successes and Failures in Technology Transfer and Diffusion*. Rome: Progetto Finalizzato Energetica, 1989.

Byrne, John and Young-Doo Wang. *Toward Sustainable Energy, Environment and Development: Institutional Strategies for Indonesia, South Korea, Malaysia and Thailand*. University of Delaware. Research report submitted to ESMAP/World Bank Alternative Energy Paths Project, August 1991.

Cambridge Energy Research Associates. *Lightening the Load: Electric Utilities and Demand Side Management*. Cambridge, MA, 1990.

Cavanaugh, Ralph C. "Least-Cost Environmental Planning Imperatives For Electric Utilities And Their Regulators." *The Harvard Environmental Law Review*, Volume 10, Number 2, 1986, pp. 299–343.

Centolella, Paul, Drury Crawley, and Edward Vine. *Energy Efficiency and the Environment: Forging the Link*. Washington, DC: American Council for an Energy-Efficient Economy, 1991.

Chalmers University of Technology, Department of Energy Conservation. "Energy Systems Technology." Annual Report 1990. Göteborg, Sweden: Chalmers University of Technology, 1991.

Chalmers University of Technology, Department of Energy Conservation. "Energy Systems Technology." Annual Report 1991. Göteborg, Sweden: Chalmers University of Technology, 1992.

Chamberlin, J.H. *Rate Design: Traditional and Innovative Approaches*. Palo Alto, CA: Electric Power Research Institute, July 1990.

Chandler, William U., ed. *Carbon Emissions Control Strategies: Case Studies in International Cooperation*. Executive Summary. Baltimore: World Wildlife Fund/The Conservation Foundation, 1990.

Chatab, I.N. "Current Energy Conservation Status in Indonesia." International Energy Conservation Symposium. Islamabad, Pakistan, 30–31 October 1988.

Chew, William and Cheryl E. Richer. "Demand-Side Management: S&P's Credit Perspective." *Standard & Poor's Creditweek* March 16, 1992.

Churchill, Anthony A. and Robert J. Saunders. *Financing of the Energy Sector in Developing Countries*. IEN Energy Series Paper No. 14. Washington, DC: The World Bank, April 1989.

Churchill, Anthony A. and Robert J. Saunders. "Global Warming and the Developing World." *Finance and Development* (June 1991).

Cicchetti, C.J. and W. Hogan. "Including Unbundled Demand-Side Options in Electric Utility Bidding Programs." *Public Utilities Fortnightly* (8 June 1989).

Contract Plans and Public Enterprise Performance. PPR Working Paper Series 118. Washington, DC: The World Bank, October 1988.

Cooperation for Development in Africa Energy Conservation Seminars: Designing and Implementing National Energy Conservation Programmes. Vol 1, *Case Study and Technical Notes*. May 1984.

Cordukes, Peter A. *A Review of Regulation of the Power Sectors in Developing Countries*. IEN Energy Series Paper No. 22. Washington, DC: The World Bank, February 1990.

Country Economics Department, Policy, Research and External Affairs, The World Bank. *The Reform of Public Sector Management: Lessons from Experience*. September 1991.

Country Economics Department, Public Sector Management & Private Sector Development Division, The World Bank. "Privatization: Lessons of Experience for Bank Group Lending." Draft Paper. Washington, DC: The World Bank, 1992.

Crousillat, Enrique. *Incorporating Risk and Uncertainty in Power System Planning.* IEN Energy Series Paper No. 17. Washington, DC: The World Bank, June 1989.

Crousillat, Enrique and Spiros Martzoukos. *Decisions Making Under Uncertainty: An Option Valuation Approach to Power Planning.* IEN Energy Series Paper No. 39. Washington, DC: The World Bank, August 1991.

Danish Ministry of Energy. *Energy 2000: A Plan of Action for Sustainable Development.* Copenhagen, 1990.

Davis, Ged R. "Energy for Planet Earth." *Scientific American* (September 1990).

Davis, Ged R. *Global Warming: The Role of Energy Efficient Technologies.* London: Shell International Petroleum Company, Ltd., October 1989.

Davis, Stacy C. and Melissa D. Morris. *Transportation Energy Data Book: Edition 12.* Oak Ridge, TN: Oak Ridge National Laboratory, 1992.

Davis, T.D. and D.R. Limaye. *Electric Utility Conservation Programs: Assessment of Implementation Experience.* Vol. 2, *Project Results.* Palo Alto, CA: Electric Power Research Institute, 1984.

Depretz, H. and P. Michel. "Improving Domestic Electric Appliances: The French Ways." *Proceedings of the* ACEEE 1990 Summer Study on Energy Efficiency in Buildings. Washington, DC: ACEEE, 1990.

Deringer, J.J., et al. "Energy and Economic Analyses in Support of Energy Conservation Standards for New Commercial Buildings in Malaysia." Lawrence Berkeley Laboratory Report LBL-23279, 1987.

Development Policy Staff. "Supporting Policy Change and the Recovery of Investment: The Inter-Relationship Between Adjustment and Investment Lending [Draft]." August 8, 1991.

Dhar, Shibu B. "Energy Conservation Workbook for Utilities." Report Submitted to U.S. Agency for International Development. Washington, DC: Energy/Development International, 1986.

Duann, Daniel J., Robert E. Burns, Douglas N. Jones, and Mark Eifert. *Competitive Bidding for Electric Generating Capacity: Application and Implementation.* Columbus, OH: The National Regulatory Research Institute, 1988.

Economic Commission for Europe. *East-West Energy Efficiency.* ECE Energy Series No. 10. New York: United Nations, 1992.

Eden, Richard. "The Greenhouse Effect: Is There a Solution?" World Electricity Conference. *Financial Times.* London, 15 Nov. 1991.

Edison Electric Institute and Electric Power Research Institute. *Demand Side Management Incentive Regulation.* Washington, DC: March 1991.

Edmonds, J.A. and D.W. Barnes. *Estimating the Marginal Cost of Reducing Global Fossil Fuel CO_2 Emissions.* PNL-SA-18361. Richland, WA: Pacific Northwest Laboratory, 1990.

Ekono. "Energy Conservation Investment Review Project." Final Report Summary. Helsinki: Ekono, 1986.

Elaah, A. and H.E. Lowitt. *The U.S. Pulp and Paper Industry: An Energy Perspective.* A Report to U.S. Department of Energy by Energetics, Inc., Columbia, MD, 1988.

Energy and Environment, 1991. Proceedings of the 1991 International Symposium on Energy and Environment. Espoo, Finland, 25–28 August 1991.

Energy Conservation Center. *89 Successful Cases of Energy Conservation*. Tokyo, 1990.

Energy Department, The World Bank. *Sector Support Strategy PaperElectric Power*. Washington, DC: The World Bank, January 1983.

Energy Department, The World Bank. *Sector Operations DirectiveElectric Power*. Washington, DC: The World Bank, May 1987.

Energy Efficiency Office. *Energy Efficiency in Domestic Electric Appliances*. Energy Efficiency Series No. 13. Her Majesty's Stationery Office, London, 1990.

Energy Efficiency Office. *Energy Use and Energy Efficiency in UK Commercial and Public Buildings up to year 2000*. Energy Efficiency Series No. 6. Her Majesty's Stationery Office, London, 1988.

Energy Efficiency Office. *Energy Use and Energy Efficiency in UK Manufacturing Industry up to year 2000*. Energy Efficiency Series No. 3. Her Majesty's Stationery Office, London, 1984.

Energy Efficiency Office. *Energy Use and Energy Efficiency in UK Transport up to the year 2010*. Energy Efficiency Series No. 10. Her Majesty's Stationery Office, London, 1989.

Energy Information Administration. *Annual Energy Outlook 1991 with Projections to 2010*. U.S. Department of Energy, Office of Energy Markets and End Use Doc. No. DOE/EIA-0383(91). Washington, DC: Government Printing Office, 1991.

Energy Information Administration. *Energy Consumption and Conservation Potential: Supporting Analysis for the National Energy Strategy*. U.S. Department of Energy, Office of Energy Markets and End Use Doc. No. SR/NES/90-02, Washington, DC: Government Printing Office, December 1990.

Energy Information Administration. *Improving Technology: Modeling Energy Futures for the National Energy Strategy*. U.S. Department of Energy Doc. No. SR/NES/90-01. Washington, DC: Government Printing Office, January 1991.

Energy Information Administration. *International Energy Outlook 1990*. U.S. Department of Energy, Office of Energy Markets and End Use Doc. No. DOE/EIA-0484(90). Washington, DC: Government Priting Office, 1990.

Energy Sector Management Assistance Programme, World Bank/UNDP. "Appendices: Compliance Guidelines for the Jamaica Energy Efficiency Building Code." Working Draft, May 1991.

Energy Sector Management Assistance Programme, World Bank/UNDP. "Kenya Energy Efficiency in the Tea Industry A Report By Consultants." 2 vols. Washington, DC: The World Bank, 1987.

Energy Sector Management Assistance Programme, World Bank/UNDP. "Kenya Urban Woodfuel Development Program Phase I Project Papers." Washington, DC: The World Bank, 1987.

Energy Sector Management Assistance Programme, World Bank/UNDP. *India Mini-Hydro Development on Irrigation Dams and Canal Drops Pre-Investment Study*. 3 vols. ESMAP Report No. 139A/91. Washington, DC: The World Bank, 1991.

Environment Department, The World Bank. *Environmental Assessment Sourcebook, Volume II: Sectoral Guidelines*. World Bank Technical Paper No. 140. Washington, DC: The World Bank, 1991.

Ernst and Whinney. *Proposals for Monitoring the Performance of Electric Utilities.* World Bank Technical Paper No. 85, Industry and Energy Series. Washington, DC: The World Bank, September 1988.

Escay, Jose R. *Summary Data Sheet of 1987 Power and Commercial Energy Statistics for 100 Developing Countries.* IEN Energy Series Paper No. 23. Washington, DC: The World Bank, March 1990.

Faiz, A., K. Sinha, M. Walsh, and A. Varma. "Automotive Air Pollution: Issues and Options for Developing Countries." Infrastructure and Urban Development Department WPS 492. Washington, DC: The World Bank, 1990.

Faiz, Asif, Christopher Weaver, Kumares Sinha, Michael Walsh, and Jose Carbajo. *Air Pollution From Motor Vehicles: Issues and Options for Developing Countries.* Washington, DC: The World Bank, 1992.

Florida Solar Energy Center. "Residential Demand Side Management for Thailand." Report prepared for IIEC, July 1991.

Foley, Gerald. "Discussion Paper on Demand Management." Eastern and Southern Africa Household Energy Planning Seminar. UNDP/World Bank Energy Sector Management Assistance Program. Harare, Zimbabwe, 1–5 Feb. 1988. Activity Completion Report No. 085/88.

Frisch, Jean-Romain, Klaus Brendow, and Robert J. Saunders. *World Energy Horizons 20002020.* Paris: Éditions Technip/World Energy Conference, 1989.

Fulkerson, W., D.B. Reister, and J.T. Miller (eds.). "End-Use Technology." *Energy Technology R&D: What Could Make a Difference?* ORNL-6541/V2/P1. Oak Ridge, TN: Oak Ridge National Laboratory, 1989.

Gadgil, A. and G.M. Januzzi. "Conservation Potential of Compact Fluorescent Lamps in India and Brazil." Lawrence Berkeley Laboratory Report LBL-27210, Rev. 1990.

Gadgil, Ashok, Art Rosenfeld, and Dariush Arasteh. "Advanced Lighting and Window Technologies for Reducing Electricity Consumption and Peak Demand: Overseas Manufacturing and Marketing Opportunities." Presented at and published in the Proceedings of the IEA/ENEL Conference on Advanced Technologies for Electric Demand-Side Management, April 4–5, 1991, Sorrento, Italy. Berkeley: Lawrence Berkeley Naitonal Laboratory, Applied Science Division, 1991.

Gadgil, A., A.H. Rosenfeld, and L. Price. *Making the Market Right for Environmentally Sound Energy-Efficient Technologies: U.S. Buildings Sector Successes That Might Work in Developing Countries and Eastern Europe.* Berkeley: Lawrence Berkeley Laboratory, Energy & Environment Division, December 1991.

Gamba, Julio R., David A. Caplin, and John J. Mulckhuyse. *Industrial Energy Rationalization in Developing Countries.* Baltimore: The Johns Hopkins University Press/The World Bank, 1986.

Gandhi, Sunita. "The Brick Industry in India: Energy Use, Tradition and Development." Ph.D. diss., University of Cambridge, England, 1986.

Gandhi, Sunita. *A Review of the Energy Situation in India.* Cambridge: Cambridge Energy Research Group, Cambridge University, 1984.

Gandhi, Sunita, Adrea Ketoff, and Jayant Sathaye. *Trends in Saturation of Refrigerators and Air Conditioners in Developing Countries: Potential for Chlorofluorocarbon Emissions.* Lawrence Berkeley Laboratory Report LBL-23745, July 1987.

Gandhi, Sunita, Florentin Krause, and Jeff Harris. *Program Design and Success: A Preliminary Overview of Lighting Programs*. Paper presented at the ACEEE Conference on Least-Cost Planning, Asilomar, CA, September 1988.

Gandhi, Sunita and Robert J. Saunders. "A Developing Country Energy Sector Response to Growing Concerns About CO_2 Emissions." Revised Draft. Energy and the Environment: Intersecting Global Issues. University of Arizona College of Law. Tucson, Arizona, 17–18 January 1992.

Gandhi, Sunita, Robert J. Saunders, Peter Cordukes, and David Ferch. "Energy Efficiency and Conservation: A Review of World Bank Experience." Draft report prepared for the Task Force on Energy Efficiency and Conservation, The World Bank, draft, 1993.

Gelil, I.A. "Egypt's National Energy Conservation Strategy." International Energy Conservation Symposium. Islamabad, Pakistan, 30–31 October 1988.

Geller, Howard S. *Efficient Electricity Use: A Development Strategy for Brazil*. Washington, DC: American Council for an Energy Efficent Economy, 1991.

Geller, Howard S. *End-Use Electricity Conservation: Options for Developing Countries*. Energy Department Paper No. 32. Washington, DC: The World Bank, October 1986.

Geller, Howard S. *The Potential for Electricity Conservation in Brazil*. Sao Paulo: CESP, 1984.

Geller, Howard S. *Promoting Electricity Conservation in Developing Countries*. Washington, DC: American Council for an Energy-Efficient Economy, 1989.

Geller, Howard S. "Residential Conservation Power Plant Study Phase 1Technical Potential." Washington, DC: American Council for an Energy-Efficient Economy, 1986.

Gellings, C., J. Chamberlin and J. Clinton, *Moving Toward Integrated Resource Planning: Understanding the Theory and Practice of Least-Cost Planning and Demand-Side Management*, EPRI, Palo Alto, CA, 1987. (EPRI document No. EM-5065, Research Project No. 2548-3.)

Gellings, C.W., V.A. Rabl, and J.H. Chamberlin. "Demand-Side Management: The Winds of Change." *Conference proceedings, Enhancing Electricity's Value to Society*, Vol. 1, Canadian Electrical Association, Toronto.

Girod, J. *Energy: Diagnosis of Energy Systems in Developing Countries*. Translated by Salm Tejpar-Dang. Luxembourg: Office for Official Publications of the European Communities, 1992.

Gocht, Werner, Klaus Knecht, Bernd Schleich, and Franz Diederich. *Solar Energy Technology Dissemination*. Proceedings of the International Seminar on Strategies and Measures for the Dissemination of Solar Energy Systems in Rural Areas of South East Asia, Manila, Philippines, 3–11 July 1989.

Goldemberg, José, Thomas B. Johansson, Amulya K.N. Reddy, and Robert H. Williams. *Energy for a Sustainable World*. Washington, DC: World Resources Institute, 1987.

Goldembery, José, Thomas B. Johansson, Amulya K.N. Reddy, and Robert H. Williams. *Energy for Development*. Washington, DC: World Resources Institute, 1987.

Greenhalgh, G. "Energy Conservation Policies." *Energy Policy* 18.3 (1990).

Griffin, Karen, et al. *1988 Conservation Report*. State of California: California Energy Commission, 1988. [photocopy]

Grubb, M.J. "Energy Efficiency and Economic Fallacies." *Energy Policy,* 18.8 (1990).

Grubb, Michael. "The Cinderella Options: A Study of Modernised Renewable Energy Technologies, Part 1: A Technical Assessment." *Energy Policy* 8.6 (1990).

Grubb, Michael. *Energy Policies and the Greenhouse Effect, Vol. 1: Policy Appraisal.* Adershot, England: Dartmouth Publishing Company/The Royal Institute of International Affairs, 1990.

Grubb, Michael. *The Greenhouse Effect: Neotiating Targets.* London: The Royal Institute of International Affairs, 1989,

Gulstone, Alfred. "Demand Management." In *Proceedings of the Regional Seminar on Electric Power System Loss Reduction in the Caribbean in Kingston, Jamaica, 3-7 July 1989,* by Olade, The World Bank, and UNDP. Washington, DC: The World Bank, 1989.

Gutierrez, Luis. *Infrastructure Maintenance in* LAC: The Costs of Neglect and Options for Improvement Vol. 2, LAC Technical Dept. Washington, DC: The World Bank, June 1992.

Gyamfi, Peter, Luis Gutierrez and Guillermo Yepes. *Infrastructure Maintenance in* LAC: The Costs of Neglect and Options for Improvement Vol. 1, LAC Technical Dept. Washington, DC: The World Bank, June 1992.

Hall, David O., H.E. Mynick, and Robert H. Williams. "Carbon Sequestration Versus Fossil Fuel Substitution: Alternative Roles for Biomass in Coping with Greenhouse Warming." PU/CEES Report No. 255. Princeton, NJ: Princeton University, Center for Energy and Environmental Studies, November 1990.

Hammond, Allen L., ed. *World Resources 1992-93: Toward Sustainable Development.* New York: Oxford University Press, 1992.

Hanson, M.E. "The Motor Vehicle Dominated City as a Non-Sustainable Urban Form: Mexico City and Jakarta." *Comput., Environmental and Urban Systems* 13 (1989).

Harvey, Hal and Bill Keepin. *Energy: From Crisis to Solution.* San Francisco: The Energy Foundation, January 1991.

Havrylyshyn, Oleh. "Independence and Economic Efficiency: A Program for Ukraine." *Transition: The Newsletter About Reforming Economies* 3.1 (1992): 1-3.

Heidell, J.A. and M.J. King. "The Industrial Sector: A Conservation Gold Mine or a Quagmire?" *Proceedings of the* ACEEE 1990 Summer Study on Energy Efficiency in Buildings, Volume 8. Washington, DC: ACEEE, 1990.

Helm, John. *Energy: Production, Consumption and Consequences.* Washington, DC: National Academy Press, 1990.

Hertzmark, Donald. "Energy Efficiency and Energy Pricing in Developing Countries." Contract report prepared for the Office of Technology Assessment, May 1990.

Hill, Lawrence J. *Energy Price Reform in Developing Countries: Issues and Options.* Oak Ridge, TN: Oak Ridge National Laboratory, August 1987.

Hirst, Eric and Carol Sabo, "Defining and Reporting Data on Utility DSM Programs," *Energy: The International Journal,* Vol. 17, No. 7, 1992.

Hirst, Eric and Carol Sabo. *Electric-Utility* DSM Programs: Terminology and Reporting Formats. Oak Ridge, TN: Oak Ridge National Laboratory and Martin Marietta, October 1991.

Hohmeyer, Olav and Richard L. Ottinger, eds. "External Environmental Costs of Electric Power: Analysis and Internalization." Proceedings of a German-American Workkshop Held at Ladenburg, FRG, 23–25 October, 1990. New York: Springer-Verlag, 1990.

Horowitz, Marvin J., Benson H. Bronfman, Gail Katz, and Philipp Degens. "Energy Check Program Evaluation." Portland, Oregon: International Energy Associates Limited, 1986.

House of Commons Energy Committee. *Energy Policy Implications of the Greenhouse Effect*, Vol. 1. Energy Committee's Sixth Report, Session 1988–89, HMSO, No. 192-I, 1989.

Igazgato, Bajusz Rezsö. *Energy and Road Transport in Hungary*. Contribution to the IRU International Study on Energy. [1983, photocopy]

Imran, Mudassar and Philip Barnes. *Energy Demand in the Developing Countries: Prospects for the Future*. World Bank Staff Commodity Working Paper No. 23. Washington, DC: The World Bank, 1990.

Ince, Martin. "Industrial Effects of UK Electricity Privatization." *Energy Policy* 16 (1988): 409-14.

Industry and Energy Department, Energy Development Division, The World Bank. *Analysis of Financing Resources for Power Development in Developing Countries, 1978–1987*. Under preparation.

Industry and Energy Department, Energy Development Division, The World Bank. *Review of Financial Covenants under Loan Agreements in the Power Sector*. Under preparation.

Industry and Energy Department, Energy Development Division, The World Bank. *Review of Forecasts for Financial Projections in World Bank SARS for Power Projects During the 1980s*. Under preparation.

Industry and Energy Department, Energy Development Division. *Review of Electricity Tariffs in Developing Countries During the 1980s*. IEN Energy Series Paper No. 32. Washington, DC: The World Bank, November 1990.

Industry and Energy Department, Energy Development Division. *The Bank's Role in the Electric Power Sector: Policies for Effective Insitutional, Regulatory, and Financial Reform*. Draft. Washington, DC: The World Bank, 1992.

Industry and Energy Department, The World Bank. *Core Report of the Electric Power Utility Efficiency Improvement Study*. IEN Energy Series Paper No. 46. Washington, DC: The World Bank, September 1991.

Industry and Energy Department, The World Bank. *FY88 Sector Review—Energy*. Washington, DC, October 1988.

Industry and Energy Department, The World Bank. *FY89 Sector Review—Energy*. Washington, DC, November 1989.

Industry and Energy Department, The World Bank. *FY90 Sector Review—Energy*. Washington, DC, October 1990.

Ingham, Alan, James Maw, and Alistair Ulph. "Testing for Barriers to Energy Conservationan Application of a Vintage Model." *Energy Journal* 12.4 (1991): 41-64.

Intermediate Technology Power Ltd. "Testing and Demonstration of Renewable Energy Technologies: Wind Technology Assessment Study Volume I, Wind Study Report." Washington, DC: UNDP/The World Bank, 1983.

International Energy Agency. *Climate Change Policy Initiatives*. Energy and the Environment Series. Paris: OECD, 1992.

International Energy Agency. *Collaboration in Energy Technology, 1987–1990*. Paris: OECD, 1992.

International Energy Agency. *Demand-Side Management: A Current and Future Resource*. Proceedings. Copenhagen, 23–24 October 1991. Paris: OECD, 1992.

International Atomic Energy Agency. *Electricity and the Environment*. Proceedings of the Senior Expert Symposium. Helsinki, Finland, 13–17 May 1991.

International Energy Agency. *Electricity End-Use Efficiency*. Paris: OECD, 1989.

International Energy Agency. *Electricity Supply in the* OECD. Paris: OECD, 1992.

International Energy Agency. *Energy and the Environment: Policy Overview*. Paris: OECD, 1990.

International Energy Agency. *Energy Conservation in* IEA Countries. Paris: OECD, 1987.

International Energy Agency. *Energy Efficiency and the Environment*. Paris: OECD, 1991.

International Energy Agency. *Energy in Non-*OECD Countries: Selected Topics 1991. Paris: OECD, 1991.

International Energy Agency. *Energy Policies of* IEA Countries: 1990 Review. Paris: OECD, 1991.

International Energy Agency. *Energy Technologies for Reducing Emissions of Greenhouse Gases*. 2 vols. Proceedings of an Experts' Seminar. Paris 12–14 April 1989.

International Energy Agency. *Fuel Efficiency of Passenger Cars*. Paris: OECD, 1991.

International Energy Agency. *Global Energy: The Changing Outlook*. Paris: OECD, 1992.

International Energy Agency. *Greenhouse Gas Emissions: The Energy Dimension*. Paris: OECD, 1991.

International Energy Agency. *Low Consumption/Low Emission Automobile: Proceedings of an Expert Panel, Rome, February 14–15, 1990*. Paris: OECD, 1991.

International Energy Agency. *Natural Gas Prospects and Policies*. Paris: OECD, 1991.

International Energy Agency. *Proceedings of the Conference on Advanced Technologies for Electric Demand-Side Management Held in Sorrento, Italy 2–5 April 1991*. 3 vols. Paris: OECD: 1991.

International Energy Agency. *Proceedings of the Workshop on Conservation Programmes for Electric Utilities Held in Paris 21–23 September 1988*. Paris: OECD, 1988.

International Energy Agency. *Substitute Fuels for Road Transport*. Paris: OECD, 1990.

International Energy Agency. *Utility Pricing and Access: Competition for Monopolies*. Paris: OECD, 1991.

International Energy Agency. *World Energy Statistics and Balances 1971–1987, 1985–1988*. Paris, 1989, 1990.

International Energy Agency and Italian Electricity Authority, *Photovoltaic Systems for Electric Utility Applications: Opportunities Critical Issues and Development Perspectives*, Proceedings of the Executive Conference, Taormina, Italy, 2–5 December, 1990.

International Institute for Energy Conservation. *Demand Side Management: Master Plan for Thailand's Electric Power System*. Final Report submitted to Electricity Generating Authority of Thailand, Metropolitan Electricity Authority, and Provincial Electricity Authority, October 1991.

Johansson, Thomas B., Birgit Bodlund, and Robert H. Williams, eds. *Electricity: Efficient End-Use and New Generation Technologies, and Their Planning Implications.* Lund, Sweden: Lund University Press, 1989.

Jones, D.W., A.P. Sanghvi, and E.L. Hillsman. *The Impacts of Inadequate Electric Supply in Developing Countries.* ORNL-6436, Oak Ridge National Laboratory, Oak Ridge, TN, 1988.

Jones, Donald W. "Urbanization and Energy Use in Energy Development." *Energy Journal* 10.4 (1989).

Jones, H. Mike, Michael Bess, and Stephen Joseph. "Energy Conservation in the Household Fuels Sector: Technology TransferObtaining Future Results from Past Experience." Proposal to the Energy Sector Management Assistance Program, The World Bank, 1990.

Joskow, Paul "Weighing Environmental Externalities: Let's Do It Right!", *The Electricity Journal,* May 1992, pp. 53–67.

Joskow, Paul and Donald Marron, "What Does a Negawatt Rally Cost?," Discussion Paper No. 596, Department of Economics, MIT, Cambridge, MA, 1991.

Joskow, Paul L. and Richard Schmalensee. *Markets for Power: An Analysis of Electrical Utility Deregulation.* Cambridge, MA: The MIT Press, 1983.

Julius, DeAnne and Afsaneh Mashayekhi. *The Economics of Natural Gas.* New York: The Oxford University Press/Oxford Institute for Energy Studies, 1990.

Kahn, Alfred E., *The Economics of Regulation: Principles and Institutions,* Volumes I & II, MIT Press, Cambridge, MA, 1988.

Kainlauri, Eino, Allan Johansson, Ilmari Kurki-Suonio, and Mildred Geshwiler, eds. *Energy and Environment 1991.* Atlanta, GA: American Society of Heating, Refrigerating and Air-Conditioning Engineers, Inc., 1991.

Kelly, Kevin, Steven Henderson, Peter Nagler, and Mark Eifert, *Some Economic Principles for Pricing Wheeled Power,* NRRI, 1987. (Publication No. NRRI-87-7.)

Kikeri, Sunita. *Bank Lending for Divestiture: A Review of Experience.* Policy, Research, and External Affairs Public Sector Management and Private Sector Development Working Paper WPS 338. Washington, DC: The World Bank, May 1990.

Kohl, Wilfrid L. *Energy and Environmental Issues: Options for U.S. Policy.* Foreign Policy Institute FPI Policy Brief. Washington, DC: The Johns Hopkins Foreign Policy Institute, 1991.

Kosmo, Mark. *Money to Burn? The High Costs of Energy Subsidies.* Washington, DC: World Resources Institute, 1987.

Kramer, Leslie. "Electrical Load Management in Malta: A Preliminary Assessment." Masters Project, University of California, Berkeley, 1988.

Krause, Florentin, Wilfrid Bach, and Jon Koomey. *Energy Policy in the Greenhouse Volume I: From Warming Fate to Warming Limit: Benchmarks for a Global Climate Convention.* El Cerrito, CA: International Project for Sustainable Energy Paths, 1989.

Krause, Florentin and Joseph Eto. *Least-Cost Utility Planning Handbook for Public Utility Commissioners.* Vol. 2, *The Demand Side: Conceptual and Methodological Issues.* Berkeley: Lawrence Berkeley Laboratory, 1988.

Kärkkäinen, S. and Y. Elmahgary, eds-. *Energy Storage Systems in Developing Countries.* Natural Resources and the Environment Series Vol. 21. London: Cassell Tycooly, 1988.

Lashof, D. and D. Tirpak. *Policy Options for Stabilizing Global ClimateReport to Congress.* U.S. EPA, Office of Policy, Planning and Evaluation, 1990.

Latin America Technical Infrastructure and Energy Division, The World Bank. "The Evolution, Situation, and Prospects of the Power Sectors in Latin America and the Caribbean Countries." Vols. I and II. Washington, DC: The World Bank, 1991.

Leach, Gerald, and Marcia Gowen. *Household Energy Handbook.* World Bank Technical Paper No. 67. Washington, DC: The World Bank, 1987.

Leach, Gerald, Lorenz Jarass, Gustav Obernair, and Lutz Hoffman. *Energy and Growth: A Comparison of 13 Industrial and Developing Countries.* London: Butterworth, 1986.

Leggett, Jeremy, ed. *Global Warming: The Greenpeace Report.* New York: Oxford University Press, 1990.

Levine, M.D., A. Gadgil, S. Meyers, J. Sathaye, J. Stafurik, and T. Wilbanks. *Energy Efficiency, Developing Nations, and Eastern Europe.* A Report to the U.S. Working Group on Global Energy Efficiency, Washington, DC, June 1991.

Levine, M.D. and X.Y. Liu. "Energy Conservation Programs in the People's Repubic of China." Lawrence Berkeley Laboratory Report LBL-29211, 1990.

Manne, Allan S. and Leo Schrattenholzer. *International Energy Workshop: Overview of Poll Responses.* Stanford University International Energy Project, July 1989.

Mashayekhi, A. "Natural Gas Supply and Demand in Less Developed Countries." *Annual Review of Energy* 13 (1988).

McKenzie, Richard B. "Sense and Nonsense of Energy Conservation." *Society* 21.3 (1992): 19-22.

McKeough, Kay. *A Study of the Transfer of Petroleum Fuels Pollution.* IEN Energy Series Paper No. 37. Washington, DC: The World Bank, July 1991.

Meier, Alan, Janice Wright, and A.H. Rosenfeld. *Supplying Energy Through Greater Efficiency: The Potential for Conservation in California's Residential Sector.* Los Angeles: University of California Press, 1983.

Meridan Corporation. *Climate Change Reference Book.* Prepared for the Office of Renewable Energy, U.S. Department of Energy, Washington, DC.

Merrow, Edward W. and Ralph F. Shangraw, Jr., with Scott H. Kleinberg, et al. *Understanding the Costs and Schedules of World Bank Supported Hydroelectric Projects.* IEN Energy Series Paper No. 31. Washington, DC: The World Bank, July 1990.

Meyers, S. "Biomass Fuels in the Developing Countries: An Overview." Lawrence Berkeley Laboratory Report LBL-27222, 1989.

Meyers, S. and J. Sathaye. "Electricity in the Developing Countries: Trends in Supply and Use Since 1970." Lawrence Berkeley Laboratory Report LBL-26166, 1988.

Meyers, S., S. Tyler, H. Geller, J. Sathaye, and L. Schipper. "Energy Efficiency and Household Electric Appliances in Developing and Newly Industrialized Countries." Lawrence Berkeley Laboratory Report LBL-29678, December 1990.

Miller, Alan S. and Irving M. Mintzer. *The Sky is the Limit: Strategies for Protecting the Ozone Layer.* Research Report No. 3. World Resources Institute, November 1986.

Mills, E. "Evaulation of European Lighting Programmes: Utilities Finance Energy Efficiency." *Energy Policy* 19.3 (1991).

Moore, Edwin and Enrique Crousillat. *Prospects for Gas-Fueled Combined-Cycle Power Generation in the Developing Countries.* IEN Energy Series Paper No. 35. Washington, DC: The World Bank, May 1991.

Moore, Edwin and George Smith. *Capital Expenditures for Electric Power in the Developing Countries in the 1990s.* IEN Energy Series Paper No. 21. Washington, DC: The World Bank, February 1990.

Munasinghe, Mohan, Joseph Gilling, and Melody Mason. *Review of World Bank Lending for Electric Power.* IEN Energy Series Paper No. 2. Washington, DC: The World Bank, March 1988.

Munasinghe, Mohan and Arun Sanghvi. *Recent Developments in the U.S. Power Sector and Their Relevance for the Developing Countries.* IEN Energy Series Paper No. 12. Washington, DC: The World Bank, February 1989.

Munasinghe, Mohan and Robert J. Saunders. *Energy Issues in the Developing World.* IEN Energy Series Paper No. 1. Washington, DC: The World Bank, February 1988.

National Academy of Sciences, National Academy of Engineering, and Institute of Medicine. *Policy Implications of Greenhouse Warming.* Washington, DC: National Academy Press, 1991.

Niculescu, I. and G. Piva, eds. *Management of Energy Resources and Energy Saving Training Programme.* 5 vols. Turin: International Centre for Advanced Technical and Vocational Training, 1984.

Noll, Scott A. "Transportation Energy Conservation in Developing Countries." Energy in Developing Country Series Discussion Paper D-73K. Washington, DC: Center for Energy Policy Research, 1982.

Norber-Bohm, Vicki. "Potential for Carbon Dioxide Emission Reductions in Buildings." Background Report for Office of Technology Assessment Climate Change Study. Cambridge, MA: Harvard University, 1989.

Nyström, Maria, et al. *Economical Stove: A study on stoves and kitchens for the Housing Area for Vinh Phu Pulp and Paper Mill workersVietnam.* 2 vols. Lund, Sweden: Lund University, 1983.

Oelert, G., F. Auer, and K. Pertz. *A Guide to Project Planning: Economic Issues of Renewable Energy Systems.* Second corrected edition. Eschborn, Germany: GTZ, 1988.

Office of Policy, Planning & Analysis, Office of Conservation & Renewable Energy. *Energy Conservation Trends: Understanding the Factors that Affact Conservation Gains in the U.S. Economy.* U.S. Department of Energy Doc. No. DOE/PE-0092. Washington, DC: Government Printing Office, September, 1989.

Office of Technology Assessment, U.S. Congress. *Changing by Degrees: Steps to Reduce Greenhouse Gases.* OTA-O-482. Washington, DC: U.S. Government Printing Office, February 1991.

Office of Technology Assessment, U.S. Congress. *Energy Efficiency of Buildings in Cities.* OTA-E-168. Washington, DC: U.S. Government Printing Office, March 1982.

Office of Technology Assessment, U.S. Congress. *Energy in Developing Countries.* OTA-E-486. Washington, DC: U.S. Government Printing Office, January 1991.

Office of Technology Assessment, U.S. Congress. *Energy Technology Choices: Shaping Our Future.* OTA-E-493. Washington, DC: U.S. Government Printing Office, July 1991.

Office of Technology Assessment, U.S. Congress. *Energy Use and the U.S. Economy.* Washington, DC: U.S. Congress, U.S. Government Printing Office, 1990.

Office of Technology Assessment, U.S. Congress. *Fueling Development: Energy Technologies for Developing Countries.* OTA-E-516. Washington, DC: U.S. Government Printing Office, April 1992.

Olson, Karen H. and Wong Wai Ching, eds. *Proceedings of the Asean Conference on Energy Conservation in Buildings in Singapore, 29–31 May 1984.* United States Agency for International Development, 1984.

O'Neill, Richard, "Competition, Efficiency and Equity in Commission Regulation," Office of Economic Policy, FERC, Washington, DC, 1990.

Ottinger, Richard L., David R. Wooley, Nicholas A. Robinson, David R. Hodas, and Susan E. Babb. *Environmental Costs of Electricity.* Report prepared for New York State Energy Research and Development Authority and the U.S. Department of Energy by Pace University Center for Environmental Legal Studies. New York: Oceana Publications, Inc., 1990.

Oxford Institute for Energy Studies. *Gas in Developing Countries.* Vol. 3, *Background Papers.* Research Project, Oxford University.

Pasha, H.A., A. Ghaus, and S. Malik. "The Economic Cost of Power Outages in the Industrial Sector of Pakistan." *Energy Economics* (October 1989).

Paul, Samuel. "The Bank's Work on Institutional Development in Sectors: Emerging Tasks and Challenges." Country Economics Department, Public Sector Management and Private Sector Development Division. Washington, DC: The World Bank, November 1991.

Pearce, David W. and R. Kerry Turner. *Economics of Natural Resources and the Environment.* Baltimore: The Johns Hopkins University Press, 1990.

Pearce, David W. and Jeremy Warford. *Environment and Economic Development: Managing Natural Resources in the Developing World.* Washinton, DC: Oxford University Press/The World Bank, forthcoming 1992.

Pearson, Peter. "Energy Transitions in Less-Developed Countries: Analytical Frameworks for Practical Understanding." Cambridge Energy Research Group, Department of Physics, Cavendish Laboratory, Cambridge. Paper No. EDP 40, July 1988.

Pertz, Klaus J., ed. *Policies for Rational Use of Energy.* New York: McGraw-Hill/GTZ, 1989.

Philips, Michael. *The Least Cost Energy Path for Developing Countries.* Washington, DC: International Institute for Energy Conservation, September 1991.

Pinto, N. and J. Besant-Jones. *Demand and Netback Values for Gas in Electricity.* World Bank Technical Paper No. 106, Industry and Energy Series. Washington, DC: The World Bank, August 1989.

Pryke, Wallace K. "Barbados Final Report on Energy Conservation." 1990. [photocopy]

Pryke, Wallace K. "Energy Conservation in Jamaica Report No. 2." London: Department of Energy, 1980.

Qazi, Azra Nuruddin. "A Technical Evaluation of Woodstoves in Rural Pakistan." Energy Discussion Paper. Edited by Ian Rudy. Cambridge: University of Cambridge, 1991.

Rau, Narayan, Mohammad Harunuzzaman, and Daniel Duann, *Uncertainties and Risks in Electric Utility Resource Planning,* NRRI, Columbus OH, 1989.

RCG/Hagler, Bailly, Inc. "Energy Conservation Investment Decisionmaking in Developing Countries: A Review of Project Implementation in Industry." Final Report. Washington, DC: Office of Energy, United States Agency for International Development, 1989.

"Reducing the Consumption of Ozone Depleting Substances in India: Phase 1: The Cost of Complying with the Montreal Protocol." Draft. London: Touche Ross & Co., 1990.

Reddy, A.K.N. *Barriers to Improvements in Energy Efficiency.* Berkeley: Lawrence Berkeley Laboratory, Energy & Environment Division, October 1991.

Replogle, M. "Sustainability: A Vital Concept for Transportation Planning and Development." Conference on the Development and Planning of Urban Transport in Developing Countries, Sao Paolo, Brazil, 1990.

Resource Management Associates, Inc. "Study of the Options for Institutionalization of the Technology Transfer for Energy Management Project in the Philippines." Final Report Prepared for The United States Agency for International Development. Madison, WI: Resource Management Associates, Inc., 1990.

Ribi, Rene, Trevor Byer, and Jean François Landeau. *Colombia: The Power Sector and the World Bank, 1970-1987.* Operations Evaluation Department Report No. 8893. Washington, DC: The World Bank, June 28, 1990.

Roberts, Paul O, et al. *Analysis of the Incremental Cost and Trade-offs Between Energy Efficiency and Physical Distribution Effectiveness in Intercity Freight Markets.* CTS Report Number 76-14. Cambridge, MA: Massachusetts Institute of Technology, 1976. [photocopy]

Rose, A.B. *Energy Intensity and Related Parameters of Selected Transportation Modes: Passenger Movements.* Oak Ridge, TN: Oak Ridge National Laboratory, 1979. [photocopy]

Ross, M. and F. Liu. "The Energy Efficiency of the Steel Industry of China." *Energy* 16.5 (1991).

Ruff, Larry, E., *The U.K. Electricity Market: The Experiment and Its Lessons,* Washington, DC, July 1991. (Available from Putman, Hayes & Barlett, Inc., 1776 "Eye" Street, NW, Washington, DC 20006).

Rupp, Petra. *The Bottom Line of Energy Management.* Manila: Philippine-German Rational Use of Energy Project of the Office of Energy Affairs, 1991.

Sanghvi, Arun and Robert Vernstrom, with John Besant-Jones. *Review and Evaluation of Historic Electricity Forecasting Experience (1960-1985).* IEN Energy Series Paper No. 18. Washington, DC: The World Bank, June 1989.

Sathaye, J., A. Ketoff, L. Schipper, and S. Lele. "An End-Use Approach to Development of Long-Term Energy Demand Scenarios for Developing Countries." Lawrence Berkeley Laboratory Report LBL-25611, 1989.

Sathaye, Jayant and Ashok Gadgil. "Agressive Cost-Effective Electricity Conservation." *Energy Policy* 20.2 (1992): 163–172.

Sathaye, Jayant, Andre Ghirardi, and Lee Schipper. "Energy Demand in Developing Countries: A Sectoral Analysis of Recent Trends." *Annual Review of Energy,* 12 (1987).

Saunders, Robert J. "Developing Countries and the Potential for Global Warmining." In *Energy, the Environment, and Public Policy: Issues for the 1990s,* edited by David L. McKee. New York: Praeger Publishers, 1991.

Saunders, Robert J. "Is It Economically Viable for Developing Countries to Cut Down on Carbon Dioxide Emissions?" *Arizona Journal of International and Comparative Law* 9.1 (1992): 205–15.

Saunders, Robert J. and Sunita Gandhi. "Alternative Paths to Energy Development: Prescriptions for the Non-Industrialized World." Washington, DC: The World Bank, forthcoming.

Saunders, Robert J. and Karl Jechoutek. "The Electric Power Sector in Developing Countries." *Energy Policy* (August 1985).

Saunders, Robert J. and Rene Moreno, Jr. "Natural Gas as a Transportation Fuel." *Natural Gas: Its Role and Potential in Economic Development*, Walter Vergara et al., eds. Boulder, CO: Westview Press, 1990.

Sayeg, Philip et al, *Assessment of Transportation Growth in Asia and Its Effects on Energy Use, the Environment, and Traffic Congestion: Case Study of Bangkok, Thailand*. The International Institute for Energy Conservation, Washington, DC, May, 1992.

Schaeffer, John, ed. *Real Goods Alternative Energy Sourcebook 1991*. Ukiah, CA: Real Goods Trading Corporation, 1991.

Schelling, Thomas C. "Some Economics of Global Warming." *American Economic Review* 82.1 (1992): 1–14.

Schipper., L. "Efficient Home Electricity Use in Indonesia." Report to World Bank, September 1988.

Schipper, L. and S. Tyler. "The Dynamics of Electricity Use in Scandinavian Households." *Energy: The International Journal* 15.10 (1990).

Schipper, Lee, Alan Meier, and Steve Meyers. "Household Electricity Use in ASEAN Countries: The Upstream Strategy for Greater Efficiency." Presented at FINESSE. The World Bank. Kuala Lumpur, Malaysia, 21–25 October 1991.

Schloss, Miguel. "Energy, Environment and Sustainable Development: Lateral Departures." Industry and Energy Division Note No. 10. Washington, DC The World Bank, Africa Technical Department, 1992.

Schloss, Miguel. "Sub-Saharan Energy Financing: The Need for a New Game Plan." Industry and Energy Division Note No. 5. Washington, DC: The World Bank, December 1990.

Schurr, Sam H. *Electricity in the American Economy*. New York: Greenwood Press, 1990.

Schurr, Sam H. and S. Sonenblum (eds.). *Electricity Use: Productive Efficiency and Economic Growth*. Palo Alto, CA: Electric Power Research Institute, 1986.

Shirley, Mary and John Nellis. *Public Enterprise Reform: The Lessons of Experience*. EDI Development Series. Washington, DC: The World Bank, 1991.

Singer, S. Fred. "Benefits of Global Warming." *Society* 21.3 (1992): 33–40.

Sioshansi, Fereidoon P. "The Myths and Facts of Energy Efficiency." *Energy Policy*. (April 1991).

Skinner, Robert. "Energy Conservation: Overview of Experiences in IEA Countries." *Proceedings of the Seminar on Energy in East and West: The Polish Case*. Paris: OECD, 1990.

Sullivan, James B. "Private Power in Developing Countries: Early Experience and a Framework for Development." *Annual Review of Energy* 15 (1990): 335–63.

Teetakeaw, P. "Energy Conservation in Thailand." International Energy Conservation Symposium. Islamabad, Pakistan, 30-31 October 1988.

Tenenbaum, Bernard and Steven Henderson, "Market-Based Pricing of Wholesale Electric Services," *The Electricity Journal*, Vol. 4 No. 10, 1991.

Teplitz-Sembitzky, W. *Basic Economics of Energy Efficiency and Energy Conservation.* IEN Energy Series Paper, forthcoming. Washington, DC: The World Bank, 1992.

Teplitz-Sembitzky, W. *Electricity Pricing: Conventional Views and New Concepts.* IEN Energy Series Paper No. 52. Washington, DC: The World Bank, March 1992.

Teplitz-Sembitzky, W. *Regulation, Deregulation, or Reregulation What is Needed in LDCs Power Sector?* IEN Energy Series Paper No. 30. Washington, DC: The World Bank, July 1990.

The World Bank. "Private Investment in Power and Coal in Asia: Selected Case Studies and Lessons." Green Cover Draft Report (ASTEG). Washington, DC: The World Bank, January 31, 1992.

The World Bank. *Managing Development: The Governance Dimension.* Discussion Paper, August 29, 1991.

The World Bank. *Manufacturers' Responses to Infrastructure Deficiencies in Nigeria.* PPR Working Paper Series 325. Washington, DC: The World Bank, December 1989.

The World Bank. *The Challenge of Development.* World Development Report 1991. Washington, DC: Oxford University Press/The World Bank, 1991.

The World Bank. *Development and the Environment.* World Development Report 1992. Washington, DC: Oxford University Press/The World Bank, 1992.

The World Bank. *Environmental Considerations in the Pulp and Paper Industry.* Washington, DC: The World Bank, 1980.

The World Bank. *The World Bank and the Environment: First Annual Report Fiscal 1990.* Washington, DC: The World Bank, 1990.

The World Resources Institute, *World Resources 1992-93.* New York: Oxford University Press, 1992.

Thurlow, George, ed. *Technological Responses to the Greenhouse Effect.* Hertfordshire, England: Rooster Books Ltd./The Watt Committee on Energy, 1990.

Train, Kenneth, *Optimal Regulation: The Economic Theory of Natural Monopoly,* MIT Press, Cambridge, MA, 1991.

Trebing, Harry M., "Realism and Relevance in Public Utility Regulation," *Journal of Economic Issues,* Vol. 8, No. 2, June 1974, pp. 209-233.

Turiel, I., R. Curtis, and M.D. Levine. "Analysis of Energy Conservation Standards for Singapore Buildings." *Energy,* 10.1 (1985).

Turiel, I., B. Lebot, S. Nadel, J. Pietsch, and L. Wethje. "Electricity End Use Demand Study for Egypt." Lawrence Berkeley Laboratory, July 1990.

UK Country Study for the Intergovernmental Panel on Climate Change, Response Strategies Working Group, and Energy and Industry Sub Group. *An Evaluation of Energy Related Greenhouse Gas Emissions and Measures to Ameliorate Them.* Department of Energy, Energy Paper No. 58, October 1989.

UNDP/The World Bank. "Energy Planning Project for European and Arab States (phase II)." *Proceedings of the Workshop on Energy Conservation in Land Transportation.* Sintra, 22-25 April 1991.

UNDP/The World Bank. "Energy Planning Project for European and Arab States (phase II)." *Proceedings of the Workshop on the Institutional & Financial Aspects for Energy Conservation Programs.* Warsaw, 15–19 October 1990.

UNDP/The World Bank. *Regional Seminar on Reducing Electric Power System Losses in Africa.* Abidjan, Côte d'Ivoire, 23-27 November 1987.

United Nations. "Energy Efficiency No. 2." *Industry and Environment* 13.2 (1990).

United Nations. *Seminar on Energy Conservation in Developing Countries Held in Stockholm 3-9 September 1989*. New York: United Nations, 1989.

United Nations. *Sustainable Energy Developments in Europe and North America*. ECE Energy Series No. 6, New York, 1991.

United Nations Environment Programme. *Energy Report Series*. Phases III, "Assessment of Tools and Methods for Incorporating the Environmental Factor into Energy Planning and Decision-Making" and Phase IV, "Comparative Assessment of the Environmental Impacts of Energy Sources." Espoo, Finland: Technical Research Centre of Finland, Espoo, 1986.

United States Energy Association. *Getting Down to Business: A Strategy for Energy Efficiency in the United States*. Washington, DC: USEA, 1992.

U.S. Congress, House. Committee on Banking, Finance and Urban Affairs. Subcommittee on International Development, Finance, Trade and Monetary Policy. *Hearing on the Implications of World Energy Trends for International Development Trade and Finance*. 102nd Cong., 1st sess., 1991.

U.S. Department of Energy. *Integrated Analysis Supporting the National Energy Strategy: Methodology, Assumptions and Results*. Technical Annex 2 of the *National Energy Strategy*, First Edition 1991/1992. Washington, DC: U.S. Government Printing Office, 1991.

U.S. Department of Energy. *National Energy Strategy*. First Edition, 1991/1992. Washington, DC: U.S. Government Printing Office, 1991.

U.S. Department of Energy. *National Energy Strategy: Executive Summary*. First Edition, 1991/1992. Washington, DC: U.S. Government Priting Office, February 1991.

U.S. Environmental Protection Agency. *Policy Options for Stabilizing Global Climate*. Draft Report to Congress, Washington, DC, 1989.

Vedavalli, Rangaswamy. *Domestic Energy Pricing Policies*. IEN Energy Series Paper No. 13. Washington, DC: The World Bank, April 1989.

Venkataswaran, S.R. and H.E. Lowitt. *The U.S. Cement Industry: An Energy Perspective*. Report to the U.S. Department of Energy by Energetics, Inc., Columbia, MD, 1988.

Vickers, John. "Economic Perspectives on Privatization." *Journal of Economic Perspectives* 5 (Spring 1991): 132.

Viklund, Elisabet. *Nordic Seminar on Domestic Energy in Developing Countries, September 1989*. Seminar Report. Lund, Sweden: Lund Centre for Habitat Studies, 1989.

Vine, Edward and Drury Crawley, Editors, *State of the Art of Energy Efficiency: Future Directions*, American Council for an Energy-Efficient Economy, Washington, DC, in cooperation with: Universitywide Energy Research Group, University of California, 1991.

Warford, Jeremy. "Environment, Growth, and Development." Report to the Development Committe No. 14. Washington, DC: The World Bank, December 1987.

Warford, Jeremy and N. van Praag. "The World Bank and the Environment: A Progress Report Fiscal 1991." Washington, DC: The World Bank, September 1991.

Weedall, Mike, Robert Weisenmiller, and Michael Shepard, eds. *Financing Energy Conservation*. Washington, DC: American Council for an Energy-Efficient Economy, 1986.

Weyman-Jones, Thomas G. *Electricity Privatisation*. Brookfield: Avebury, 1989.

Williams, Robert H. "Low Cost Strategies for Coping with CO_2 Emission Limits." Critique of "CO_2 Emisison Limits: and Economic Cost Analysis for the USA," by Alan Manne and Richard Richels. *The Energy Journal* 11.3 (1990).

Williams, Robert H., Eric D. Larson, and Marc H. Ross. "Materials, Affluence, and Industrial Energy Use." *Ann. Rev. Energy 1987* 12:99-144 (1987).

Yates, Philip. *Improving Power System Efficiency in the Developing Countries Through Performance Contracting.* IEN Energy Series Paper No. 4. Washington, DC: The World Bank, May 1988.